MODERN DAIRY GOATS

No. 2
COUNTRY BOOKS

No. 2 COUNTRY BOOKS No. 2

General Editor BRIAN VESEY-FITZGERALD

MODERN DAIRY GOATS

by

MARY DOUGLAS GORDON

LONDON

NICHOLSON AND WATSON

First published in 1950 by
Ivor Nicholson & Watson, Ltd.,
London, W.1

Printed and Bound by
Love & Malcomson, Ltd.,
REDHILL, Surrey

CONTENTS

INTRODUCTION

The growing interest in small livestock is ample justification for such a treatise as this upon goat breeding and management.

The author, an old friend and one-time colleague of mine, is well versed in her subject and is well known for her writings and broadcasts on goats, poultry, and other stock.

Her present textbook covers all aspects of goat culture, from breeding to feeding, together with the care and treatment of the milk and the making of butter and cheese. In short, this book embraces the whole subject of goat keeping, from A to Z.

Mary Douglas Gordon writes from a thoroughly practical knowledge of this craft and draws upon a lifetime's experience of goat culture in presenting her detailed information, which will be, both to the amateur and to the more experienced, a veritable vade-mecum.

R. E. Louch, N.D.D.

Lecturer in Poultry Husbandry, Cheshire School of Agriculture; Cheshire County Poultry Instructor; and late Lecturer in Poultry Husbandry and Dairying to the Holmes Chapel Agricultural College.

Cheshire School of Agriculture,
Reaseheath,
Nantwich,
Cheshire.

CHAPTER ONE

WHY NOT KEEP A GOAT ?

ARE you thinking of keeping a goat ? Do you think it would be very nice to have your own milk, butter, and cheese but fear the business of keeping a goat or two would be beyond you ? If you feel that way about it, think again. Do not let yourself be fobbed off too easily with the idea that you could not manage it, for it is probably much easier than you think. Many people who want to keep goats could easily do so, if only they knew it. It does not even matter if you have no pasture, for many goats have none at all and give excellent results. And, as for experience, you gain that as you go along and find it surprisingly easy.

Of course, you could not keep a goat in a town flat,—unless you had an allotment where she could live. But you do not necessarily have to be in the heart of the country. Goats are accommodating creatures, and one finds them kept, and giving heavy yields, in the most unexpected places. If you have a place where you could keep a few laying hens, then you have a place where you could keep a goat. And, as regards feeding, they are most catholic in their tastes and are fed in very much the same way as rabbits and are as little trouble.

Perhaps you already have a pen of poultry and a few rabbits. Probably you have a dog and perhaps, also, a cat. Then why not a goat? " Ah! " you may say, " a goat is a different proposition altogether. Anyone can look after hens and rabbits, but I don't know anything about farm stock." But this is where you are mistaken, for a goat is no more trouble than a few rabbits and certainly less trouble than poultry. All you have to do is to feed, and water, and milk her, twice a day, and clean her place out once or twice a week, just as you do with rabbits. If you can tether her out to graze, so much the better, as it will save you the trouble of giving so much food in the stall ; but many goats are kept by people who are away at work all day and attend to them in the mornings and evenings, just as they do with their rabbits and poultry.

Perhaps you are wondering in the back of your mind, " What about kidding ? " Do not let that worry you. The average goat has her young as easily and quietly as a cat has kittens. In thirty years of livestock breeding, I have never had to assist a goat in this natural function, and the times I have had a vet. on my premises can be counted on my fingers and have only been due to accidents and malicious poisoning. So you need not worry about that part of the business. Just think of the trouble you take to rear youngsters for the poultry pen !

Actually, many goats are looked after almost entirely by the children of the family, and excellent little goatkeepers they make. Tots of four and five become expert milkers, and goats are seldom awkward with children as they sometimes are with adults. My own little daughter looked after my goats entirely when there was illness in the house, and, finding a newly-kidded nanny one morning, she promptly rubbed down the kids, put them in the shopping basket, and staggered upstairs to show them to me ! So do not let thoughts of difficulty or inexperience prevent you from keeping a goat or two. And think of the great benefit of having an ample supply of delicious, wholesome milk, produced on your own premises, with enough left over to make some butter and cheese!

There is just one bogey that must be laid forthwith. One still hears people asking if goats' milk is really wholesome, and if it has not an unpleasant " goaty " taste. Old legends die hard, and this belief is a relic of the days when the only goats to be seen were poor little " scrubs,"—small, shaggy, horned creatures, of no pedigree and often of decidedly uncertain temper. They were generally left out in all weathers and got nothing to eat but what they could find, and their owners were content with the odd pint or two, which was all they could be expected to give. No wonder the milk was "strong", as I freely admit it was,—and still is, if you keep scrubs.

Actually, those days are not so very far behind us, for it was the two world wars that brought goats to the fore and changed them beyond recognition. Rationing, and the many shortages and difficulties of those two hard wars and the years that followed, forced people to try to produce at home what they could not get in adequate quantities through the usual channels, and goat breeders rose valiantly to the occasion. The modern dairy goat is a large, shapely, affectionate animal, either naturally hornless or disbudded within a few days of birth, so that the horns do not grow. She may give you as much as two gallons of milk a day, at her best. A six-pinter is now considered quite mediocre. And she will continue in milk for two, or even three, years, without being re-mated. She does *not* "smell". The stud male does, but you do not need to keep one. And her milk does not taste "goaty", except when it is stale, and the flavour can then be removed by boiling the milk and allowing it to cool. This is no different from cows' milk, which gets distinctly "cowy" when stale, unless it has been pasteurised, when it goes, not stale, but rotten. Moral: Use your milk while it is fresh!

In this connection, I will tell you a little story which is perfectly true. At one of the pre-war Dairy Shows which I reported, a certain lady wandered into the goat section and was amazed to find the goats were large and beautiful animals, quite unlike the poor, smelly little scrubs she had seen somewhere in the country. She stayed and watched the large pails of milk being filled and talked with one of the stall-holders about the goats and was given some leaflets. She agreed that goats' milk might have many virtues but said it had an unmistakable flavour and that she simply could not get used to it. After a good deal of persuasion, she agreed to try a glass of newly-milked goats' milk and give her unprejudiced opinion. " But I know I shan't like it ! " she said. The stall-holder sent off an assistant, who duly returned with a large glass of milk. The sceptical one sipped it, screwed up her face, sipped a little more, said it was not so bad, and eventually finished the lot. " Well, it really wasn't at all bad. In fact, I quite enjoyed it," she admitted. " But, you know, there *is* a flavour. Very slight, I admit, but you can't quite get away from it." The stall-holder then broke the news that the milk she had just drunk had come from the Jersey cow section,—especially fetched for her! She took it in good part, but it just shows what prejudice can do, doesn't it ?

This book, then, is written for the men and women who start from scratch, knowing nothing about goats and perhaps even having prejudices to overcome. It is written in a way you can understand, even if you have never seen a goat. Yet it will, I hope, be helpful, also, to those who have some experience. It will tell you what to expect in a goat and how to avoid being " sold a pup ", what the different breeds are like, how to feed and manage them, how to treat them in illness, and, better still, how to keep them well. Goats are, by nature, healthy animals, and a sick one is an unusual sight. They are clean by nature and will not eat dirty food or drink stale water ; and no doubt that has much to do with their milk being so good and wholesome that doctors recommend it for babies and invalids, who can keep nothing else down, and prescribe it for stomach ulcers, that painful and trying complaint so widespread after years of wartime diet.

CHAPTER TWO

BREEDS AND STRAINS

THERE are several different breeds now vying for public favour, all of them good milkers ; and, if your only aim is a regular milk supply, there is not much to choose between them. In fact, one often finds people making their choice on æsthetic grounds only ; and that is no bad thing, for you do best with an animal you like and admire. In appearance, they differ a good deal, and some people are greatly taken by the looks and colour of a certain breed, while others do not like it at all. Some people, for instance, have a distinct fancy for white animals, while others prefer them coloured. Again, you may find yourself attracted or repelled by the unusual style and " Roman nose " of the Anglo-Nubian ; or you may love the striped face and silky tassels of the Swiss-type goats.

Whether you just want goats for milk, or whether you are seriously taking up breeding and exhibiting, I do strongly advise you to see as many goats as possible, before making your choice, and to choose what appeals to you æsthetically. And the best way to do this is to attend livestock shows. You can find out when and where these are being held by studying your local papers, or by applying to the British Goat Society, and you will doubtless gain a good deal of useful information while enjoying a pleasant day's outing.

The following descriptions and illustrations will give you an idea of what the popular breeds really look like:—

The British Toggenburg. This breed has been evolved through the crossing of Swiss Toggenburgs with British goats, and it is an improvement on the original. It is a large goat and a heavy milker, though the butter-fat percentage is not always as good as it might be. The colour is most attractive, being a warm fawn shade, (or sometimes chocolate), with white or cream markings. The markings consist of a facial stripe from above the eyes to the muzzle, with light edges and tips of the ears to match, also light legs from the knees downwards and a little more of the white or cream on the rump. It is a most attractive patterning and sets off the body colour to great advantage.

The ears are carried erect, and the facial line is either straight or what is known as "dished". This means there is a slight drop between the eyes and the end of the muzzle, giving a concave effect. These goats often have two little silky tassels hanging from the fronts of their necks.

The British Alpine. This fine milking breed has, also, been evolved by British breeders, but it has no Swiss counterpart, in spite of its name. It is of the Swiss type, with the same white, cream, or light fawn markings, but on a black body colour. They, also, have erect ears, a straight or dished facial line, and often the same silky tassels. In fact, taking the above detailed description of the British Toggenburg, and reading "black", instead of "fawn", for the body colour, you will have a very good idea of the appearance of the British Alpine.

The British Saanen. This breed, like the British Toggenburg, has been evolved by British breeders, by crossing the Swiss Saanens with British goats, and, again, it is a great improvement on the original. It is a large goat, generally white but sometimes cream, and the breed is noted for its heavy milkers. The Swiss type is retained, with the erect ears, straight or dished facial line, and often the silky tassels.

These three breeds are of the same type,—the Swiss type,—though larger and heavier than their pure Swiss ancestors. But there is a breed of goat much admired in this country that is of a different style altogether. This is

The Anglo-Nubian. These goats are so very different in appearance from the prick-eared goats that they are sometimes mistaken, at first sight, for large dogs. They have long, drooping, pendulous ears, like a spaniel, and the facial line is convex. This gives the " Roman nose " effect, which, with the long, hanging ears, and often bizarre coats, makes them exceedingly striking. The coat is short, fine, and glossy, like that of a pony, and any colour is correct. Often one sees the most remarkable mottled and marbled effects. Marked or self-coloured coats of any colour are in order. They look like the native goats one sees in Biblical and Eastern pictures, and they are, in fact, the result of crossing the pure Nubian with English goats a good many years ago. The pure Nubians could not stand our climate, and, though Anglo-Nubians are sturdy enough for most places, they should not be kept in very severe conditions. These goats are not usually such heavy milkers as those of the Swiss type, but their milk is very rich in butter-fat, and they are known as " The Jerseys of the goat world ". This is a point worth noting, if you want to make a lot of butter, or if you want the richest possible milk for the rearing of valuable puppies or kittens.

British Goats. Under this heading come all goats, other than English, that are not eligible to be entered on the registers of any of the breeds already given or those of their Swiss prototypes. They are usually, though not always, of Swiss type, and they can be any colour, and there are some excellent milkers among them. If you have, say, a British Alpine and an Anglo-Nubian and you send them both for service to a breeder who has a British Saanen at stud, you are mating pure-bred, pedigree stock on both sides. Perhaps all these goats are registered and have fine pedigrees and records of performance behind them. But, because you have crossed two different breeds, you cannot enter the off-spring on the registers of pure-breds. The kids, no matter what their type and colour, will be known as British.

English. These goats are seldom seen nowadays. They are small, rough-coated animals, with horns, and have not been bred for milk production. Their only interest is to the few breeders who do not want to see the native goat of England die out altogether.

This brief summary of the breeds, together with the illustrations given, should enable you to get a clear picture in your mind's eye of the popular goats of to-day. They have changed out of all recognition, even during my own lifetime. The years between the two world wars have been used by far-sighted breeders to turn what was once an unattractive, and often actively unpleasant, little animal into a large, beautiful, well-balanced, affectionate creature. The high-yielding goats of about twenty-five years ago were often pathetic-looking specimens. Breeding for high records in milk production, without corresponding body development, often resulted in tiny little creatures carrying enormous, ungainly udders that looked as though they did not belong to the animal and often trailed on the ground and were trodden on by the goat.

To have gone on in this way would certainly have led to a long and weary trail of disease and lack of stamina, as it has done with other domestic livestock, when high records have been pursued regardless of body development. But goat breeders, backed up by the British Goat Society, took a longer view of things. All the breeds previously mentioned, with the exception of English, have been carefully bred and selected for many generations, with a view to improving milk yield, butter-fat percentage, and size and type of goat. Careful records have been kept by the owners of specially good strains, the milk being weighed after milking. These records, and the goats themselves, have to be available for the inspection of the British Goat Society's officials, and goats cannot be registered, or entered in the Herd Book or Foundation Book, unless they have certain qualifications and reach a certain standard. This means that there has been a steady and continual improvement in the yield and type of goat. There are certain very outstanding goats, but this is not the only thing. The general *average* of production has risen to the point where the average yield of twenty-five years ago strikes us as decidedly poor.

The World Record is, at present, held by a six-year-old British goat, Malpas Melba, which gave 5,993 lbs. 6 ozs. of milk in the officially-recorded year, 1946-47. Melba holds, also, the Lactation Record for 365 days, (from four days after kidding), with a yield of 6,661 lbs. 3 ozs.

The World Record for 1945-46 was held by a British Saanen, Bitterne Joke, with a recorded yield of 5,928 lbs. Her yield for the 365 days, (commencing four days after kidding), was 6,422 lbs. 3 ozs.

Previous to this, the World Record had been held, since 1937, by another British Saanen, the well-known Hartye of Weald, with a recorded yield of 5,479 lbs. 1 oz. Hartye was a beautiful specimen and an example of how a good milker can, also, be a good looker. Her consistent holding of the World Record for nine years is equalled by her amazing list of triumphs in the exhibition world. In addition to winning twelve " Firsts " in Inspection Classes at County, Royal Agricultural, and Dairy Shows, she gained twelve British Goat Society Silver Spoons for " Best in Show ", the Holmes Pegler Trophy, and numerous other Cups and Trophies, and her blood has had a powerful influence on the breed.

It will be noted that milk yields are reckoned in pounds. It may sound strange to the novice, but this is always done, both with goats and cows ; and it makes it easier to understand if you bear in mind that a pint is approximately twenty ounces, and a gallon is about equal to ten pounds.

These certainly are outstanding goats, but high production is now more the rule than the exception, as will be seen from the following excerpt from the British Goat Society's regulations regarding entry into the Foundation Book:—

"Alternatively, a goat is eligible, irrespective of pedigree, if it qualifies for the R2 prefix, (i.e., yields not less than 2,000 lbs. of milk, officially recorded, in any recorded year ending October 1st), or a "Star" or "Q Star", (i.e., an affix granted to goats that obtain a minimum number of points in a milking competition at a recognised show,—usually regarded as the equivalent of a gallon milker)." Now, the goats entered in the Foundation Book are by no means all champions. They represent a fair average of what the modern dairy goat can do. And, when we realise that 2,000 lbs. of milk a year works out at nearly six pints a day, (without allowing for any dry period prior to kidding), it will be seen that the *average* is now very good indeed. And it is the average that is the concern of most of us.

You should now have a pretty good idea both of what modern dairy goats look like and what you can expect of them ; and the next thing to do is to see them in the flesh, before deciding which breed to choose. You can do this, either by attending agricultural shows and country fêtes or by going to see reliable breeders. Under the Stud Goat Scheme, run by the British Goat Society under the auspices of the Ministry of Agriculture and Fisheries, there are now pedigree males of high quality at stud in most of the counties of England and in some parts of Wales ; and it is easy enough to get the addresses of these breeders and make arrangements to go and see their goats. Do not be in too much of a hurry to commit yourself. See several herds, and, if possible, all breeds, before you make your final decision, for there are few things so pathetic as discovering you like another breed better than the one you have and yet, at the same time, being unwilling to part with the animals you have, because you have become attached to them. For, make no mistake about it, you can become attached to goats, just as you become attached to dogs. They are far more individual than cows and have distinct personalities. Also, if you have children, you are almost certain to find they make pets of the goats. It is by no means unusual to find a goat ambling round the kitchen, looking for titbits and gazing at you in dignified surprise if you remark upon it! In fact, I have seen my goats stroll into the dining-room and take half a loaf off the bread board!

All the breeds I have mentioned, with the exception of English, have been bred for high production. There are outstanding specimens in all these breeds, and averages are good as well. But there are, also, poor goats and indifferent ones going under these names, and there are even pure-bred, pedigree goats that one would not call satisfactory. To understand this, one must consider the question of Strain. By this is meant certain blood-lines within a certain breed. For instance, Mr. A. and Miss B. may both keep pedigree British Alpines, but Miss B.'s Alpines may be very heavy milkers while Mr. A.'s may be mediocre. Perhaps Miss B. has been at it longer and has put many years of careful work into evolving her heavy-milking strain. Perhaps Mr. A. is a comparative beginner, or perhaps he is satisfied with a lower average and does not take his goat breeding too seriously. Again, Mrs. C. may become sentimentally attached to her goats and be unable to cull an animal that she knows ought not to be kept. Whatever the reason, the result is that some strains are much better than others and that you must consider strain as much as breed when making a purchase.

There is one thing you must avoid at all costs and that is buying from dealers who buy up cheap goats all over the country, run them with a billy, and sell them off as anything they seem to resemble. Some of these people run really good billies, and so there is the possibility of some of the kids being reasonably good; but these gentry seldom know anything about the nannies they buy. Often they are semi-wild. Invariably they are poor producers, otherwise they would hardly be for sale in the open market. There is such a heavy demand for goats nowadays that good prices are readily obtainable for them ; and the really heavy milkers are seldom for sale, for the owner of a two-galloner knows its worth and prefers to keep it at home and breed from it.

If you see an advertisement offering goats at ridiculously low prices and advising you to write and " state wants ", go warily. Few genuine breeders keep all breeds. If you buy cheap animals from doubtful sources, you will almost certainly get a poor quality animal of no pedigree, and it is not even certain that it will be in kid. Most breeders keep careful notes of the mating dates, especially when they run a large herd and sell stock, for it is unsatisfactory, to say the least of it, to be unable to say for certain if, and when, a nanny has received service.

You will, of course, find genuine breeders who keep more than one breed. Again, you will find many breeders who ring the changes with regard to stud males, keeping one of a different breed each season and letting him go at the end of the breeding period. This is done to avoid having to mate back the young nannies to their own father, and people who adopt this practice will have a pleasing variety of goats to offer. But these will be British. Even though not pure-bred, they will be pedigree stock, and the owners will be able to give you the necessary particulars as to parentage and milk yields, and so you will know what you are getting. The trouble with the people who buy up anything cheap, and shove it in a field with a billy and hope for the best, is that they do not know themselves what they are handling.

CHAPTER THREE

HOW TO CHOOSE A GOAT

THIS chapter is going to begin by answering some very elementary questions regarding terms used by goat keepers, so perhaps you experts would like to skip it ! If you do read it, however, do not do so with a superior smirk ; and do not, I beg of you, bestow the superior smirk on people who ask you the same questions. Believe it or not, each of these questions has been asked me by a normal, sensible adult ; and, childish though some of them may seem to us who are used to animals, we must remember that townspeople often just do not know the answers. And doubtless they are experts on other subjects on which we should ask equally childish questions. Curiosity in a novice is to be encouraged and rewarded with helpful information. And, to the beginner, I say, " Ask, and keep on asking. It is the best way to learn. And, if you are not satisfied with the answers you get, try somebody else. Anyway, keep on asking ! "

What are milch goats ? This term simply means goats kept for milk production. The Americans just say " milk goats "; but here, in Great Britain, we still use the term that dates back to Old Testament times.

The proper term for a male goat is a buck, and a female is called a doe. But, colloquially, they are known as billy goats and nanny goats, respectively. The young of goats are called kids, while they are under a year old. A young nanny, between the ages of one year and two years, is known as a goatling.

" A first kidder " is a young nanny that has produced, or will shortly produce, her first set of kids ; and " A second kidder " is a nanny that has produced two families, and so on. The term has *no* reference to the order of birth of the different kids in one set. All the kids produced from one pregnancy are called a set, not a litter, as with rabbits. One or two kids at a time are most usual, though threes and fours are by no means uncommon.

A disbudded kid is one that has had the horn buds treated with caustic potash or a proprietary disbudding stick when a few days old. This prevents the horns growing. Great efforts are being made by breeders to produce naturally hornless goats, and no male can be admitted to the Stud Goat Scheme unless it is naturally hornless. But many of the female goats that carry no horns were intended by Nature to have them but were disbudded as kids. The words " dis-horned " and " de-horned " are sometimes used, especially in American publications.

Reference to a goat's " bag " means her udder, and the former word is more frequently used. One speaks of a goat " bagging-up " or " making a bag ", when her udder enlarges and begins to fill, prior to kidding.

Do not be shy of the word " belly ", which is freely used among livestock keepers and, indeed, must be, if confusion is to be avoided. If you say an animal should have a white belly, you mean just that. You must not politely substitute the word " stomach ", because, when we talk of stomachs, we mean the organ of that name, in which digestion takes place. In the case of the goat, there are four,—the rumen, the recticulum, the omasum, and the abomasum.

If a goat, or, more likely, a kid, is said to be " pot-bellied ", it means it is blown up with indigestion. It is the whole belly area that is inflated, not just one of the stomachs. There are, also, various local names for this condition, such as " blows ", " pod ", " hove ", and " hoven ".

The tassels at the throats of some goats have no connection with milk production. They are just appendages, generally much admired for their pretty effect; but whether or not a goat has them is of no consequence.

In some parts of the country, a goat that is not in-kid,—that is to say unmated,—is sometimes called barren. One might presume that a barren animal was one that would not breed, but that is not necessarily the case, when speaking of goats. It simply means that she is producing nothing. She has gone dry and she has not been served by a billy. Incidentally, this is bad husbandry. A goat should be re-mated before she goes dry. Then she will continue to give milk during most of her pregnancy.

Now that you know the meaning of these various terms, let us consider what your first purchase shall be. Will you start with a young kid and rear it ? Or will you have a goat in milk, or a mated goatling ? If the latter, shall it be a first or second kidder, or would an older goat suit you best ? In every case, there are advantages and disadvantages ; so let us consider them carefully.

If you start with a goat in milk, you will have milk straightaway, and this is an obvious advantage. The only thing is it will cost you more than a kid or a goatling. Still, if you want milk, this should not worry you unduly, for to buy the milk would cost plenty. And actually, at the time of writing this book, it is impossible to obtain more than the miserable ration of two-and-a-half pints per week, unless you can get goats' milk ; so the cost of goats has soared to exceptionally high figures, and a really heavy-milking goat costs as much as a pre-war cow.

Before the second world war, it used to be estimated that the value of a milking goat was one pound sterling for each pound of milk given when at the peak of her lactation. Thus, a nanny giving a gallon a day, when at her best, would be worth round about £8. With goats giving more than ten pounds of milk a day, the value would be assessed at half as much again, or £1 10s. 0d. per pound of milk. At the time of writing, prices are much higher than this. A gallon milker will easily fetch £12 or £15, and many heavy-milking goats change hands at £20 and more. Doubtless prices will settle down again to a more reasonable level, when rationing and the shortages of war are things of the past. At present, one can only give the above figures as guides.

Of course, good milking goats can still be obtained for less than this. Anything from £5 to £10 is a reasonable price to pay for a reasonably good goat. Breeders who keep records and are always striving for higher yields consider a six-pinter not worth keeping; and, without such breeders, goats would never have reached their present high standard. But, for the man or woman who just wants to keep goats so as to have plenty of milk available, the six-pinter is by no means to be despised. Quite possibly, a moderate milker may suit you better than a heavy one, for there is a limit to how much milk an ordinary family can use. Beware, however, of a goat in milk offered you at less than £5. It *might*, of course, be all right; but there is probably some snag about it.

Except for the high initial cost, a goat in milk is a better "buy" than a kid or goatling. Being adult, she is fully grown and developed, and you can see for yourself what sort of an animal she is. You can find out at once how much milk she is giving, and you can ask for past records of her previous performance—how much she gives at her peak, how long she remains in profitable production, and so on. The length of lactation,—i.e., the time a goat remains in profitable production,—is an important factor. A really well-bred milking goat will keep on milking for two years, or even more, without being re-mated, and you can arrange kiddings so that you always have a goat in milk. But goats that only keep up for a few months and have to be re-mated every year are poor investments.

If you decide to buy an adult goat, you have the choice of a dry goat in kid, an in-kid goat still giving a little milk, a newly-kidded goat in full milk, a goat partly through her lactation, or a dry, unmated goat.

The last-named should not be considered, unless you already have goats and are making an addition to your herd. A dry, unmated goat will give you nothing until you have got her mated and kept her through her pregnancy. A dry goat in kid is a better proposition, but, even then, you have to keep her for nothing until she has her young ones ; and it is not every beginner who wants to start with a kidding and the problems of what to do with the babies and how to milk a newly-kidded nanny. Some of them are very difficult at this time and need an expert hand to get them settled.

It is best to have a goat in milk, and, if her yield has dropped and she is towards the end of her lactation, it is best to get the breeder to have her served before you take her. You will then be able to get your hand in at milking, and have the advantage of an immediate supply, while knowing that your goat is probably pregnant and so will not be dry on your hands for long. Breeders do not usually guarantee a mated goat to be actually in kid, unless the time is so advanced that the condition is obvious.

A newly-kidded goat in full milk is a good choice, if you are already an expert milker and are in need of a heavy supply. Goats travel quite well within about a week of kidding, but it is best not to have them from such a distance that they are a long time on rail. After a journey, the yield often drops ; but it should pick up again, when the goat has settled and got used to her new quarters and new owner. A complete novice, who has to learn how to milk, should not choose a goat in full milk, for they are often restive and difficult when their udders are very distended, and they are quick to sense when a hand lacks confidence. However much she may need the relief of being milked, she may "play you up" and hold her milk from you. You may give up in exasperation, when you have got a certain amount. Or you may persevere, in spite of her antics, until you think you have milked her dry, when, in fact, you have not. In either case, the result will be that the milk yield will steadily drop, for Nature adjusts the supply to meet demand, and you must strip the animal quite dry every time you milk, if you expect to keep up a good yield. It is, therefore, safer for the novice, who must learn how to milk, to do so on a goat more advanced in lactation, as she will be gradually dropping, in any case, and is, also, less likely to be troublesome than one with a very full udder.

As regards age, no goat does her best during her first lactation. She gradually improves with each kidding until she reaches her third, when she should be at her best. After this, you do not usually get higher yields; and, when the goat is five or six years old, her lactations begin to get poorer each time, until she is no longer worth her keep. Cases have been recorded of goats of ten and twelve years old, and even more, continuing to breed and give a pint or two a day, and this is often enough for a cottager. In fact, it is a good deal more than our allotted ration of the war years. But such old goats need careful feeding, as their teeth are generally so worn that they cannot graze.

Initial expense can be saved by starting with a kid or a goatling, but, in either case, you will have to spend quite a lot before you see any return. So, unless you already have goats and are merely increasing your herd, a kid or a goatling is not a good "buy". A kid must have milk. You can introduce part calf meal when the kid is a month old, but a certain amount of milk is needed for a time, and, in any case, the rearing is going to cost you a good deal, in one way and another. It will remain to be seen what sort of an animal your kid makes, and then you cannot have her mated before fifteen months, at the earliest. So it will be two years before you can have your kid in profit.

A goatling is a better proposition than a kid, because you can then see what sort of a goat she will make, and it will not be so long before she comes into milk. At the same time, you have no proof that the goatling will be a good milker. You can only judge by her size and development and the records of her ancestors. Also, a first kidder is often very nervy and troublesome and has to be broken in, so to speak, as regards being milked. So, considering one thing and another, it is best for the beginner to start with adult goats in milk, in the first instance. When you have got used to milking and general management, you can then try your hand at kid-rearing and bringing up young goatlings from your own stock. In this way, you gain your experience gradually and do not have too many new problems on your hands at once.

Before you actually make a purchase, you should know how to judge age and quality. The age of a goat is found by looking at its teeth. Like sheep and cows, goats have no incisors (or cutting teeth) in the upper jaw. Their place is taken by a hard pad of gristle. There are eight incisors in the lower jaw and twenty-four molars, (or grinding teeth), in all, the molars being in both upper and lower jaws. It is by looking at the incisors in the lower jaw that the age is judged. When a kid is a year old, it should have its full complement of thirty-two teeth, but the eight incisors will be small, white, and pointed. Early in the second year, the two middle incisors should drop out and be replaced by two permanent incisors, and these are much larger and more square in shape. In the third year, the two baby teeth next to the big permanent incisors are changed; in the fourth year, two more are changed; and, in the fifth year, the last and outermost of the incisors are replaced by permanent ones. The goat is then said to be "full-mouthed".

After this, one can only judge age by the state of the teeth, since no more are changed. The molars show signs of wear, as the animal grows older, and the incisors get discoloured. Eventually, the latter begin to fall out, and then you know the goat really is on the down-grade. An aged goat, that has lost most of its incisors, cannot graze properly.

It is difficult to say just when the first permanent incisors will fall out, as a good deal depends on the kind of life the goat has led. Animals that go out and do a lot of rough browsing, barking trees and gnawing off branches, are much harder on their teeth than those which get plenty of hay and concentrates. Probably seven to eight years is an average age for losing the first permanent incisors. But one must use one's judgment and consider how much wear and tear the teeth are likely to have had in the life the goat has been leading. In this respect, the condition of the coat may help you to form an opinion, for a goat that goes out a good deal and "roughs it" has a thicker, coarser coat than one that spends a good deal of time in its stall. Each case must be considered individually, whether it is an adult goat or a kid, for there are always exceptions to the general rule, and growth and development vary.

To judge quality in a milch goat, one should consider size, build and general conformation, texture of coat, and the udder. A heavy milker must have size. She needs a good spring of ribs, to accommodate a strong, large heart and lungs; she needs a roomy body, in which to digest the large amount of food she must eat to produce plenty of milk; and she should stand firmly, with hind legs placed well apart, to take the weight of her large body and to accommodate a big, well-filled udder.

A good milker should be wedge-shaped, whether viewed from behind or from the side. It is a bad fault if a milker looks narrow from behind, with legs too

A good milker is wedge-shaped when seen from the rear. Note wide stance, feet firmly planted to take weight of the big frame and to accommodate large udder.

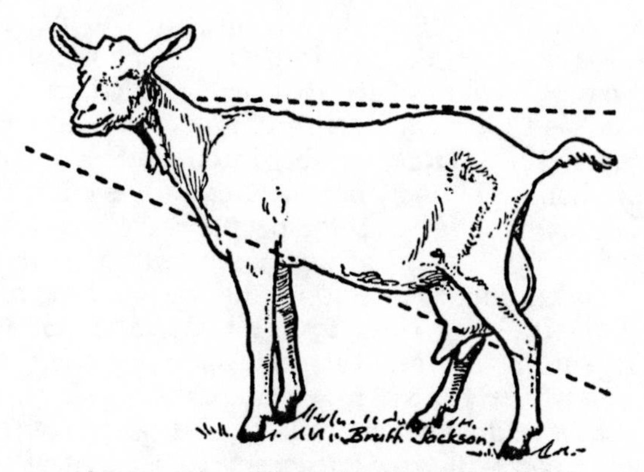

Note wedge-shaped body of good milker, slender neck, wide stance of hind legs, large udder, and good teats.

close together. Viewed from the side, she should be much deeper at the hindquarters than at the shoulder. An animal that carries much weight in front is a meat animal, not a heavy milker. This same principle applies to other farm animals and, also, to poultry. For

Beefy type—too heavy in front for a milker. Note thick, masculine neck, poor udder, and tiny teats.

meat, you want weight in front ; but, for production, you want weight in the hinderparts. A masculine-looking nanny is not likely to prove a good milker. Concentrate on the neat, feminine type, with rather a small, narrow face, large, bright, intelligent eyes, and fairly fine bone. Do not, however, carry the craze for fineness of bone too far, as some people do. This leads to lack of stamina. Just avoid masculinity and beefiness. The skin should be soft and pliable, and the coat should be soft, lying close to the body, and not too long. Again, in this latter respect, one must use one's judgment and be guided by circumstances. A hairy, shaggy goat is seldom a good milker ; but a fine-haired, well-bred milker will grow a thicker, coarser coat, if she goes out a good deal, than she will if kept much in her stall. She will not, however, grow very hairy and shaggy.

The udder is very important and needs a paragraph to itself. It should be soft and silky to the touch, and, after milking, it should appear to shrink up a good deal. The animal that seems to have almost as big an udder after milking as before is not a good milker. She has a fleshy udder. Coarse, fleshy udders and very hairy udders are not found on heavy producers. Bear this in mind, if you are buying a goat near the end of her lactation. If she really is a good milker, her udder will probably appear quite small, and this may seem deceptive ; but you can generally form a pretty sound opinion by feeling it and noting the quality. Also, the teats will still be of good size and placed well apart. Tiny little teats that are hard to grasp are a great drawback, especially when a man has to do the milking.

Do not worry if the goat appears thin, so long as she is not unduly so and has a bright eye and alert expression. A goat that is a heavy milker is always on the thin side, for she turns much of her food into milk, instead of putting it on her back, as meat animals do. Goats are, by nature, somewhat ungainly animals, with protruding hip-bones ; and people who are not used to them are sometimes inclined to pity a goat for being thin, when she is actually in excellent condition. A fat goat should always be avoided, not only because she is not likely to be a good milker but, also, because an overfat animal often fails to breed. A certain amount of excess flesh can, however, be expected in well-grown kids and goatlings, for they are still growing and have not come into production. When they do, they soon settle down to what will be their normal appearance.

CHAPTER FOUR

HOUSING

It is very nice to see one's goats housed in a neat row of proper goat stalls, all built to plan, but it is more than likely that your first goats will have to make do with a converted stable, tool-shed, or poultry house, or something of that sort. Actually, many goat keepers never do have buildings specially made for goats but just keep them in whatever is convenient; and there is no reason why they should not.

Goats are accommodating creatures and do not ask for super quarters. The fine ranges of stalls, etc., one sees, from time to time, are, as often as not, built to please the æsthetic taste of the owners! If you have plenty of money to spare and goats are your pleasure and your hobby, you can spend a good deal on making their premises a pleasure to the eye; and a certain amount of uniformity in style does save labour and add to the convenience of the attendant. But the goats themselves do just as well in quite primitive quarters, so long as they are dry, free from draughts, and well ventilated. These are the three essentials.

Goats, being creatures of the mountains, do not take kindly to fugginess. They cannot bear being shut up as closely as cows are in winter. Actual draughts are, of course, as bad for them as for any other animals, and direct currents of air across the bodies of tethered goats must be avoided. But the goat must have an abundance of fresh air at all times of the year, and arrangements for this must be made. The fact that it is cold air does not matter. Goats suffer far more from heat than from cold. Lowness of temperature does not matter, so long as it is not very extreme. So it is always best to have the stable type of door, with the bottom portion to fasten independently of the top. The upper part can then be left open during the day on almost every day of the year, and often it can be left open at night as well.

If you are converting a shed with a single door, this can easily be altered, by sawing it across and adding two more hinges and another fastener. Add another staple or button, (according to the type of fastener you are using), to the wall of the house, so that the upper portion of the door can be secured when open. This will prevent the door being damaged through swinging about in windy weather.

It goes without saying, of course, that the walls and roof must be weathertight. If you are going to use a really old building, it may need some repairing. A few slats may be required to cover cracks in the walls, and perhaps the roof may need re-felting. The floor must not be damp by reason of low position, but the dampness caused by urine is a different matter altogether, and suitable litter will prevent it making the whole floor wet. The goat must have a dry sleeping place.

As regards floors, some people like them of wood and some of concrete, while others are quite satisfied with earth floors. An earth floor is perfectly all right, so long as it is well littered and you do not have sickness among your goats. But, if disease comes, it is very difficult to deal satisfactorily with an earth floor, so that it shall be safe for fresh animals.

Wooden or concrete floors have the great advantage that they can be washed down regularly. They can be thoroughly scrubbed between kiddings and disinfected when necessary. There are no regulations regarding drainage, in connection with goats, as there are with cows, and it is not incumbent on the goat-keeper to make arrangements for swilling down his goat sheds; but it is obviously to his advantage to be able to do so, since cleanliness of the goats and their surroundings improves the keeping properties of the milk. The desirability of scrubbing out the sheds between kiddings is recognised by the Government, which allows a supplementary soap ration for the purpose.

It is sometimes said that concrete floors are cold and may tend to encourage rheumatism, especially when goats are kept a good deal in their stalls. This may well be the case if the floors are insufficiently littered, or if unsuitable litter is used, so that the goats can scratch it away and lie on the bare floor ; but the writer has used them a good deal and has never experienced any such trouble. And concrete floors have the advantage that they are rat-proof and do not harbour insect pests. This is an advantage, because, where there are livestock, food will be dropped ; and, where food is dropped, rats and mice will seek to gain access ; and these vermin bring with them insect pests and disease germs.

Goats are very fond of pushing away the litter and lying on the bare floor, doubtless because, in their natural state, they are used to the bare mountainside and rocky ledges. They do not like to be bedded down on litter, especially when it is soiled, and there is much to be said for their cleanly tastes. If you must use litter, a good choice is a layer of peat moss covered by a liberal allowance of straw. The peat moss absorbs the urine and makes a firm bed that it is difficult for the goats to push away, and the straw above keeps clean much longer than if it is used alone.

The average goat has a passion for getting up on to something hard and bare. Some only get as far as pawing away the bedding and sleeping on the bare floor, and this, of course, is dangerous. But, if they can possibly manage to get up on to something, you may be sure they will do it. I have even known goats get up into their mangers and fold up their long legs in the most incredible manner and curl up and go to sleep. It really is a ludicrous sight to see a matronly figure with a well-filled udder arising out of a narrow manger ! It may be that this is done sometimes to avoid floor draughts, which all goats detest. A draught

BRITISH ALPINE GOAT
Didgemere Dulcie. The property of Mrs. Arthur Abbey.

BRITISH ALPINE MALE
Broomfield Duke. The property of T. A. Urie.

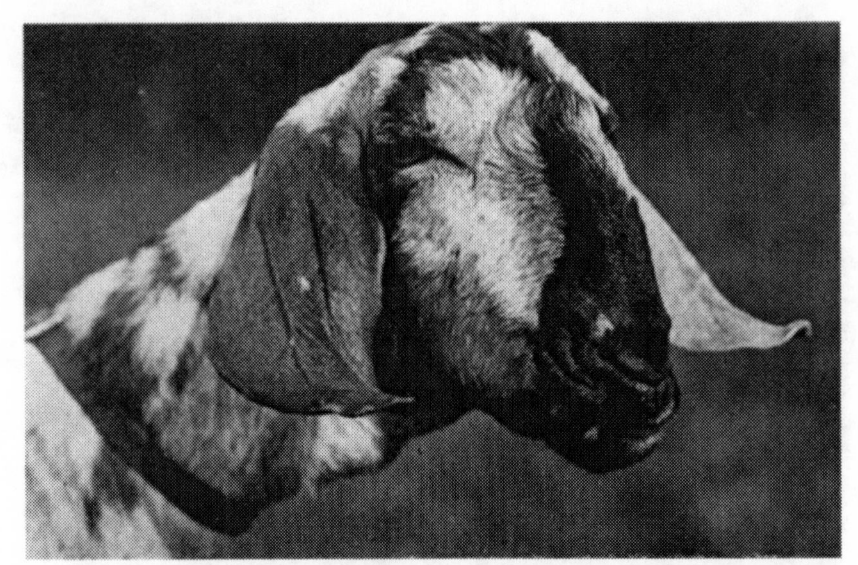

FINE HEAD STUDY OF ANGLO-NUBIAN GOAT
Theydon Barbarette, Q*. The property of Miss K. Pelly.
[*"Sport and General" photograph.*

STRIKING ANGLO-NUBIANS
Vanitye and Vivette. The property of Miss K. Pelly.

is so very different from open air all round, as in the natural state. I have, however, known goats that persistently sought for sleeping places above floor level when there certainly were no draughts, and I think it is due to some dim ancestral memory of rocky ledges.

Be this as it may, goats do like to get up on things, and you will keep them happy and save yourself much expense on litter if you make your goats sleeping platforms. These are just like shelves on legs, and they can be taken out and washed whenever necessary. They may be made of planking or slats, but, if the latter, see they are not too far apart, as the goat's feet are very small. I have always used planking, but, if I used slats, I would not put them more than half an inch apart.

Any man or woman who can use a hammer, nails and saw should be able to make these sleeping platforms, and, once made, they last indefinitely. There is nothing about them to wear out or get out of order; and, if soiled or infected, they can easily be cleansed. First, you need to make a frame of 2 in. by 2 in. timber, setting it on legs of the same material, about twelve inches high. This makes a firm and steady base, and the boards or slats are nailed across. The platforms do not need to be very large, for goats love to draw themselves into a small, compact heap when sleeping,—another relic of ancestral memory, I believe, when a loosely-laid body might topple off its favourite ledge! A small goat would find ample room on a shelf or platform 1½ ft. by 2½ ft., and 2 ft. by 3 ft. would accommodate a large goat.

There should be some windows in the goat house, so that there will be light and ventilation if the door has to be shut. The kind of window used does not matter a great deal, so long as it fulfils its purpose, but personally I prefer hopper windows, like those used in poultry houses. As they fall inwards from the top, they admit fresh air without draught. Also, they can be removed in an instant for washing, since they only slide in, and

HOPPER-TYPE WINDOWS
(seen from inside)

The wings at each side are of galvanised zinc. One side is bent up, for fixing to the woodwork with two screws. The other side is bent the other way, to take the weight of the glass, when open. The glass is not fixed, but can slide in and out.

they can be left out altogether in very hot weather. Another advantage is that there is no fixing or puttying to be done, if you have the misfortune to break a pane. You just buy another piece of the correct size and slip it into place.

Whether special goat-houses are built, or you make use of existing sheds, I strongly advise you not to have too many fancy gadgets. They are only so much more for you to keep in order—and so much more for the goat to amuse herself by bashing about when she is bored. Goats can be very hard on their houses and "furniture", even if they are hornless, and odds and ends sticking out here and there are a standing tempta-

tion. I have seen goats deliberately whacking their heads against their hayracks,—just for the fun of seeing them come down,—and then look round with an expression that said: "There you are! Another nice little job for you!" Their heads seem able to stand an indefinite amount of whacking and are their favourite tools for "doing housework". Another favourite little dodge is heading a bucket over and then thudding on its base with the front feet, until it is well dented!

Personally, I think it is a waste of good buckets to give them to goats. Each animal should have its manger or a food box made of heavy material and fastened to the wall. It should, also, have its hayrack and holders for salt and mineral licks. These, with a staple or rod, to take the chain, and a sleeping platform, are all that are needed, unless you have your goats in pairs, or more, and so need partitions.

Hayracks of metal can be bought and fixed to the wall, or they can be made at home of timber. I have some excellent ones, made by an old soldier, of ordinary wood cut from a spinney. The bark was first peeled off, and the sticks were left in the sun to dry. If this had not been done, the hayracks would have been mistaken for dessert and treated accordingly! When the wood was hard and dry, it was trimmed into rods, about 1 in. in diameter and 15 ins. long. These were nailed, top and bottom, to a long, flat cross-piece, 16 ins. by 1½ ins. Two exactly similar pieces were taken and nailed across in exactly similar fashion on the other side, one across the top and one across the bottom, so that the ends of the rods were well protected. The rods themselves were 2 ins. apart.

This looked like a miniature fence, with its twin cross-pieces at top and bottom. It was easily made, and it was then ready for fixing. This was done by driving two long nails through the twin cross-pieces at the base and through into the wall of the house. They were not driven right home, so that there should be enough play for pulling out the top of the rack to the required

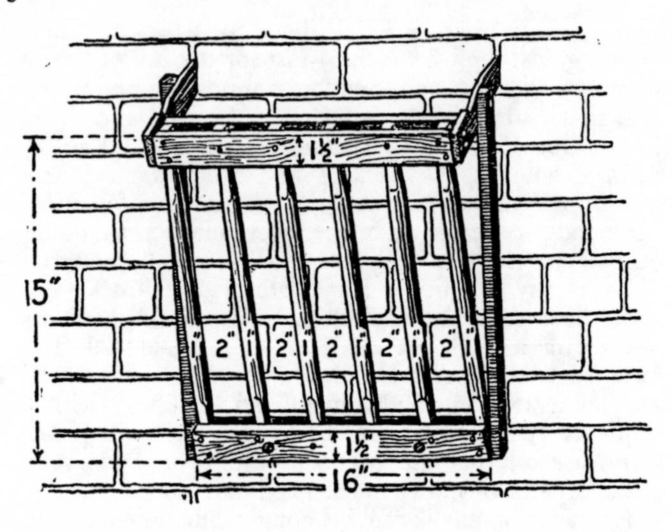

angle. A wooden peg was then taken and driven into the mortar between the bricks at the top right-hand corner of the rack and another at the top left-hand corner. The top of the rack was drawn out, (using the play allowed by not driving the bottom nails right home), and the ends of the wooden pegs were nailed or wired to the corners of the rack. All that was needed then were two small strips of wood, slightly longer than the rods in the frame, to nail down either side, to prevent hay being pulled out in large quantities. Hay was put in at the top and pulled out through the bars, and these little home-made racks prevented much waste. If the walls of the house had been of wood, it would have been necessary to nail a piece of 1 in. by 1½ ins. across, on a level with the top of the rack, and nail the ends of the wooden pegs into the ends of this piece, at right angles. In fact, I have done this with brick walls, when I could not find suitable places for driving in the pegs.

It is well worth turning one's hand to jobs of this

sort, for the ready-made articles can run away with a lot of money. I have had these racks in use for years, and they are still as good as ever, except for a small amount of gnawing done by kids. They are easy to make and do not call for a high degree of skill, and they can be made at odd moments, or on rainy days, and put aside for use as required.

Galvanised metal holders for salt and mineral bricks can be bought quite cheaply and hung from nails on the wall beneath the hayracks. These are excellent while the licks, (which are made in brick form), are new; but, when they have been hollowed out with licking, they often break and fall out, and then I put them into the bottoms of the hayracks and the goats continue to lick them through the bars. Lumps of rock salt can also be placed at the ends of the mangers.

If you are running your goats in pairs,—either one pair in a small shed or many pairs in a proper goat-house,—you will either have to have partitions or tether them at some distance from each other. The partition does not have to be the whole length of the stall. It is only a case of preventing the goats turning round and soiling the litter where their stall-mates lie. These partitions are, however, rather difficult to fix, and, unless you are yourself an expert, it is best to have a local carpenter to do them for you. They must be strongly made and fixed, as they come in for a good deal of bashing and rubbing.

The orthodox measurements for a single goat-stall are 2 ft. by 2 ft. or 2½ ft. by 2½ ft., according to the size of goat. But why make a single stall? While you are about it, you might just as well make double stalls. Goats like company of their own kind and, as a rule, do better with stall-mates. Sometimes one finds a goat that likes to be by herself; but most of them like a pal, and many form firm friendships and insist on keeping the same pal, which is sometimes a little inconvenient! I remember the time when I bought a goat, guaranteed quiet and docile, which nearly drove us frantic, until I wired for her stall-mate. Then all was peace and quietness!

This, of course, was an exceptional case, but, even so, there is not much point in making single stalls or building single goat-sheds, even if you only plan to have one goat. It is quite likely you will decide later to have another, so that you have no break in the milk supply; and, if you already have room for her, trouble and expense will be saved. And, of course, one double goat-house costs less than two single ones, since less material is needed.

Some people are greatly against tethering goats in their sheds and like to have a big shed, after the style of a barn or loose-box; but I do not recommend this system for adult goats. However careful one may be to keep horned and hornless goats apart, there is always a certain amount of teasing and bullying in an untethered herd, whether it be two goats or twenty. There is, also, always a "boss" in the herd. And, if the goats are left loose in a shed, there is the risk of injury, especially with in-kid nannies and those heavy in milk.

A large, open shed of this sort is, however, very useful for growing youngsters, especially during the winter months. For this purpose, it is best to have as few gadgets as possible, or they will certainly be used as playthings and probably broken. By all means provide something strong and steady, on which they can play, for this will help to keep them happy and healthy, —but it need not be your fittings! It is a good idea to have a few stalls, so that the young animals can shelter in them when the temperature is low at night. Anything in the nature of an alcove will attract them, and you may be surprised at the number of kids that will curl up together in such a place. They keep themselves warm in this way, and, so long as the building itself is well ventilated, they will not lack fresh air through clustering together in stalls or corners. I have known suitable alcoves made by curtaining off corners with sacking, but this is definitely only a temporary measure, as they soon tear up the sacks.

In such a building, a number of hayracks should be provided in different places, otherwise the stronger will keep the weaker and small ones away from the food that is always the mainstay. A crib in the centre of the building, with access on all four sides, such as is used for cattle, is a good idea ; and one goat-keeper of my acquaintance has, for years, made a point of picking up old-fashioned iron bedsteads at sales, for this purpose. Normally, they go for a mere trifle. I once saw over a hundred go for a shilling each. With a couple of yards of sheep fencing on either side, attached to the rails that run at head and foot of the bed, you have an outsize hayrack that can hold enough to stay a number of goats throughout the day, and the hay can be pulled through on all four sides. Of course, they do not look exactly ornamental, but they are cheap, strong, and serve their purpose, and they can be used indoors or out-of-doors.

If you are embarking on goats on a fairly large scale, you will probably need a loose-box for your stud male ; and the same sort of places are useful for kidding goats, which are best left untethered in a place of their own. Some people let them kid in their stalls, as cows do, regardless of the other animals beside them, just taking away the kids, (and sometimes the mother), when they are born. But this is not to be recommended. It is kinder to let a kidding goat have a place to herself for a short time ; and, if there should be any unfortunate developments, you will be thankful you kept her away from her fellows and so have the trouble under your control.

Before we leave the subject of housing, there is just one other system to be mentioned; and, for cheapness and simplicity, it has everything else beaten to a frazzle. Actually, it is only meant for use during the warm months of the year, but I have seen it followed all the year round in the southern counties of England. Still, one must remember that it is only meant to be a system of camping out for the summer. Anyone who kept goats in this way, and had no alternative accommodation, would indeed be in a quandary in a winter like that of 1946-7.

Movable field shelter, with only half of front boarded and no door. Note that the chain is only long enough for the goat to lie down in the shelter, and is not long enough for the goat to round and overturn it. There should be a swivel hook on the end of the chain.

The system simply amounts to supplying each tethered goat with a light, movable shelter, which is moved along with her each day, when her pin is moved. It has sides, roof, and half a front, but no door and no floor. It is placed with its back to the prevailing wind, and it must be put at the very end of the goat's reach, so that she can get in and lie down but not go round it and overturn it. I have seen such shelters made of roofing felt on a light wooden frame, but these do not last very long, as the goats damage the felt by poking at it with their sharp hooves and sometimes even nibble it. Made of wood on quite a strong frame, they are still light enough to be moved by hand by one person, and they give years of service.

The goats stay out all day and night, using their shelters as they please. Some goats will stand to be milked, and this saves a lot of trouble. You just go down with the food, and water, and milk the goat and move her, and you do not have to bring her up from the field. You lose some of the benefits of the system if you have to bring them up twice a day to the stalls for milking. But you still save on litter, and the goat is free to graze at any time during the twenty-four hours. It is, also, a useful system for young goatlings and goats that are not in milk. If you have ample pasture, goats kept in this way will find their own keep, so long as they have enough range and are moved often enough. Very heavy milkers will need something extra, but it is surprising how much a goat will find during the months of lush growth, so long as she has time and opportunity.

Some systems of goat-keeping necessitate a yard connected with the goat sheds. If you have little or no pasture, you must have some place into which you can turn the goats while their sheds are being cleaned, and a yard fenced with cleft chestnut paling serves the purpose very well. However, this will be dealt with in more detail in the next chapter.

CHAPTER FIVE

THE USE AND ABUSE OF PASTURE

If you have pasture, or if you have access to free land on which goats can be pastured, (such as commons, strays, or wide road-side verges), you will be able to keep down your food bill. But, if you only have a very limited amount of pasture, it must be used with care and discretion. Indiscriminate use by goats of a very limited area,—or, for that matter, overstocking of any area, whatever the size,—results in the land becoming goat-sick. When this happens, the animals become unthrifty and eventually sicken, and losses will be very heavy, unless you get them off the goat-sick land before the illness has a real grip of them. It is difficult to cure this complaint and quite impossible unless you get the goats away from the cause of the trouble ; so remember this, if you think of taking over a place on which a previous tenant has recently kept goats. Find out how many he kept, where he kept them, and if they remained healthy and profitable.

This disease, which is a form of auto-poisoning, is unknown among stall-kept goats, and many goats are kept in perfect health and production with no pasture at all. Many a suburban goat lives in a converted tool-shed, with a little yard attached in which she airs herself while her sleeping place is cleaned. Some goat owners take their goats out for walks on chains, just like dogs, allowing them to stop here and there, to nibble bits of this and that. Many prize goats are kept in this way, as the prejudice against tethering is very strong among breeders, and not all of them have facilities for giving their goats free range.

If you have a certain amount of pasture but not enough, my advice to you is to combine the systems of stall and pasture, so that the land is used to your advantage but not abused. The goats could be put out during certain hours of the day and then brought up to their stalls. I have followed this system for years, with excellent results, and I find the goats are often pleased to come in and lie down and cud after a morning's grazing. You will have to arrange your actual system according to the amount of land you have, and perhaps, also, to fit in with your own plans and other work. My system is to give the goats hay first thing in the morning and let them have their concentrates when they have been milked. After we have had our own breakfast, the goats go out to graze, coming in again about tea-time, as a rule. Sometimes they come in before mid-day and go out again when the heat is less intense, as the summer sun often distresses them, if there is no natural shade.

You may not be able to put your goats out for as long as this, but that need not worry you. The main thing is to prevent your pasture from becoming goat-sick; and you can make your own plans with this end in view, allowing the goats to be out just as much as is wise and then bringing them in and giving them food in their stalls or yards. If your land is very limited, you may find it best to use it only during the spring and summer, keeping the goats up during the cold months and stall-feeding them entirely, while the grass-land recovers. In such a case, a dressing of lime, at the rate of 10 cwt. to the acre, given at the end of the grazing season, would help recovery.

Of course, if you have ample pasture, or have access to an unlimited amount of free grazing, problems of this sort need not worry you; but always be careful to see no particular part of the land is over-used, either by preference on the part of the goats or by careless management on your own part. Do not keep them long in one field, if you have others available. And, if you give hay and roots out-of-doors, do not feed them always in the same place, just because it is convenient.

We now come to that vexed question of tethering.

Many of the well-known breeders are strongly opposed to it, and certainly it can make a goat's life a misery. Anyone who has seen a poor little scrub tethered on the roadside on an ordinary dog chain, scouring its little circle to the roots in search of food, and having no shelter from the elements, cannot fail to be sorry for the poor little thing. It is cruel, and it is, also, short-sighted, for no goat kept in that way can give of its best. But there is tethering and tethering; and many people would be unable to keep goats at all, if the use of the chain and pin were abolished.

There are two cases where tethering is absolutely essential, and the first is that which we have just been considering,—i.e., where the amount of pasture available is very limited. Goats are not careful grazers. In fact, they are not grazers at all, in the strict sense of the word. They are browsers. They do not go placidly along the field, like a cow, munching steadily and eating down the grass as they go. They like to eat a bit here and a bit there, running about from place to place and often trampling down more than they eat. Even a goat on a chain will, if she is well fed in her stall, waste a good deal when the growth is lush, by trampling down good food merely to reach some tiny titbit that she fancies. So, if you must conserve your grassland, you will have to use tethering pins and take your goats methodically up and down the field, so that they do not return too soon to any one place.

The second case where tethering is essential applies to the vast majority of goat-keepers. It is when you have no goat-proof boundaries. And, believe me, there is precious little that is goat-proof except a high brick wall! And, even then, it really would have to be a high one, for their leaping powers are truly incredible. The average goat-keeper probably has a grass field or two, surrounded by the usual hedges, which may or may not be in good condition. If he can run to it, he may supplement these with chestnut paling or sheep fencing, but this means a considerable outlay.

You only need to keep goats for a matter of weeks to find out how easily they can get through the average hedge. Not only are they marvellous leapers, but they can stretch themselves up and pull down branches you would swear they could never even touch. If the mouth cannot do it, the front hooves will be called into play; and, if anything more is needed to give an extra inch or two, out comes a long tongue and laps itself round the nearest twig, and down comes the whole branch, to be held securely with a forefoot. And these branches may be poisonous evergreens, hanging over your neighbour's wall. Or they may be his pet roses!

In addition to this incredible genius for stretching and leaping, a goat can also wriggle through tiny little weak patches that one would hardly think big enough for a terrier. They can pack themselves up small and they can stretch themselves like snakes; and you may put your goat in a nice paddock with plenty of food, thinking the hedge is sound and good, and find her, ten minutes later, streaking up the main road or polishing off the last of somebody else's hardy annuals! So, what with one thing and another, most goat-keepers fall back on tethering, as a means of keeping their goats both out of mischief and out of danger.

Of course, there is always electric fencing, if your operations warrant it, and if you have enough land. With this system, you purchase an Electric Fencing Unit, which runs on a battery and sends an intermittent current along wires used to enclose the chosen area. It is harmless to both goats and their owners, but touching it gives a very unpleasant sensation and animals soon learn to respect it. Quite light stakes can be used, as no pressure will be put on the wires. Three strands are needed for goats, since they are liable to try to jump over or wriggle under a single strand, which latter is sufficient for cattle. The top strand should be placed thirty-six inches from the ground, the second should be twenty-four inches, and the bottom strand should only be twelve inches from ground level.

The bottom strand should not be electrified, unless you are running young kids loose in the enclosure ; and, if it is electrified, you must be careful to keep the grass short beneath it, and see no bushes or brambles touch it, otherwise there may be a short circuit.

This system is gaining favour rapidly, for it is easy to erect and remove, as required, and saves the stock-owner much worry. Also, there is a saving in stakes, since they need not be very stout or high and they can generally be put as much as twelve yards apart. However, since the majority of goat-keepers have few goats and little land, they generally fall back on tethering.

HOW TO TETHER GOATS

If you have to adopt this system, give your goat as much range, within reason, as is possible ; and always make use of any natural shade available, when the weather is very hot or there is a cold wind blowing. In addition to this, take precautions to see that a tethered goat cannot get herself into a position from which she cannot recover herself. For instance, do not tether her near a deep dyke or sudden drop of any sort, or she might slip down, or even climb down after some fancied titbit, and then find herself unable to get up again and so strangle in her own collar, held back by the weight of her own body. Goats are naturally very agile and nimble, but you must always remember that a tied goat has not the same freedom of recovery as one that is loose. An untethered goat that slips over a sudden drop may lose her footing and roll on to her back, but she will soon be on her feet again and scrambling up the bank, unless she is actually injured.

Another thing to beware, when tethering goats, is the presence of shrubs, and saplings, and anything of that nature. If the goat, in her wanderings, gets her chain round a stout stem, she may not have the sense to unwind herself, and the results are likely to be fatal. Some goats will just stand patiently, knowing they are caught up, and waiting for someone to come and unwind them. Some will actually bleat for help. But others just lose their heads and twist themselves up tighter than ever, in their frenzy to escape. I remember seeing a lovely Toggenburg goatling strangled in this way. It was only quite a little sloe sapling, barely two feet high, round which she had caught her chain; but, being tough and green, it had merely bent with her chain and become so hopelessly entangled that it had to be cut away in bits. If it had snapped, the goat could have got free.

Now, as regards the actual tethering, the best way is to use a chain and a wire cable. The wire is securely pegged down at either end, and the goat's chain has a ring attached to one end, which runs along the cable. In this way, the goat has a very wide range and will not need moving more than once a day. Also, it is very unlikely that she will be able to get herself free and go trotting off into danger and mischief.

If you must use a chain alone, let it be a reasonably long one. The ordinary dog chain is not sufficient, unless you are able to move your goats several times a day or only put them out for short periods. A goat that is to be tethered out all day on a chain without a cable should have a length of three yards. The chain will have a spring fastener, for attaching to the collar, and the other end of the chain goes through a loop of iron at the head of the tethering-pin and goes back a foot or so and is slipped through a ring on the chain itself.

These rings on the chains need watching, for goats are amazingly strong and often the rings become elongated and eventually snap. If you find one of your rings getting out of shape, cease using it and knot the end of the chain securely to the ring on the tethering-pin. The rings on the pins are much stronger, being made of iron by the blacksmith when he makes the pins.

Tethering pins,—and, in fact, all goat equipment,—can be bought ready-made, but it is cheaper to get such heavy stuff locally and most country blacksmiths will make them to your order. There are various styles, but, if your land is light, your pin should be not less than 2½ ft. in length and should have a corkscrew end, to enable it to get a firmer grip in the soil. With heavy land, a shorter pin, or even a wooden stake, may do. A good deal depends on the goats themselves, and you will find which are the placid ones that are content to "stay put" and which of them imagine any place to be better than where you have put them! I have had the most persistent escapers, and some used their brains to a considerable extent. One nanny would walk steadily round in circles, leaning into her collar as hard as she could, until eventually she had enlarged the hole so much that she could easily draw up her pin and stroll off on her own. Unfortunately for her, she preferred the road to the soft grass and made such a noise clanking her chain and pin along that she always got caught and taken back again in next to no time! Most of my escapers made a bee-line for home, and they would come straight up to the front door and knock loudly with their heads and hooves. As soon as the door was opened, in came the visitor!

If you leave the blacksmith to make the pins according to his fancy, you will probably find he puts a strong iron ring through a hole in the head of the pin. These work well enough, but it is better to get him to make a neck an inch or two below the head of the pin and make a double loop in the form of an S, one end of which swivels round in the neck of the pin. This means that the chain never gets twisted round the pin, no matter what the goat does, and this is a distinct advantage. With a plain loop through a hole in the

A MODEL GOAT BYRE

The property of Messrs. Gregory, Ltd., Liverpool. Here is the answer to people who ask if goats can be kept commercially.

FINE WINTER QUARTERS FOR THE GROWERS

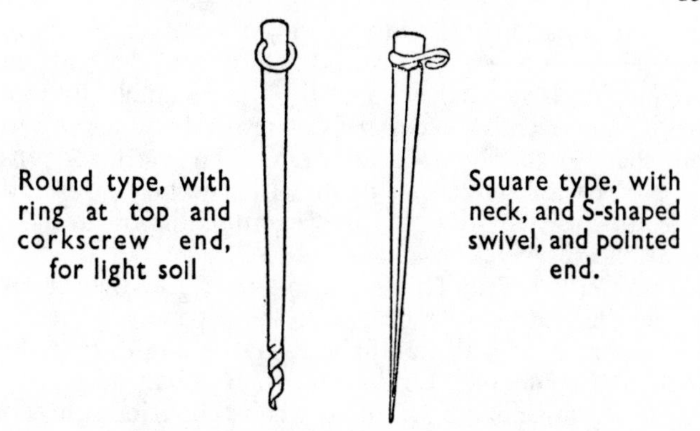

head of the pin, one often finds a goat completely tied up, patiently waiting for someone to unwind her. I have never known a goat strangled in this way, as they do when the middle part of the chain gets twisted round a sapling. I think the goat realises that there is safety at the head of her tethering-pin and that it is anxiety to get back to the familiar pin that makes them lose their heads, when their chains are caught up by something else. Still, even though no tragedy may result, it is a pity to have goats twisted up short, if it can be avoided, since they are not getting their grazing. Pins with necks are liable to snap when you are driving them into the ground ; but, for that matter, pins with loops through holes in the head are liable to split across the top. They all go eventually, and you have to take them back to the blacksmith to be heated and repaired, so it is best to keep one or two spares. They are quite cheap. Even during the war years, I was never charged more than one and sixpence each.

Strictly speaking, the tethering-pin should be driven in flush with the ground. This does away with the risk of twisting up and shortening of the chain. When the land is wet, or if your soil is very light, even a 2½ ft. pin, driven in right up to the head, can easily be pulled out by a determined goat, and it is essential to have the pins long. But these long pins are difficult to drive right in when the ground is hard and dry. Much hammering is needed, which all helps to crack the top, and it may even be necessary to use the hammer round the sides, to get them out at night. So one finds pins are not always driven in flush with the ground at all times of the year,—hence the recommendation to have an S-shaped swivel.

The chain is usually hooked to a strong leather collar round the neck of the goat, but some people prefer to use leather head-stalls, and these have a ring to take the hook of the chain. Undoubtedly, a collar does spoil the appearance of a goat, for the constant friction wears away the hair. The ridge, or fringe of stiff hair that runs down the neck and along the backbone, is spoilt by having broken hair where the collar rubs, and a persistent puller often becomes quite bare under the neck. If you want to show your goats, or if you just do not like to see the appearance spoilt in this way, get head-stalls for your goats. They are rather expensive in the first place, but they last many years, if carefully treated. Remember that leather needs feeding, and give them an oiling when they look dry. Also, polish them with brown shoe polish.

An adult male goat, kept for stud, should always wear a head-stall, for their strength is prodigious and you have little control over a goat held only by something round the neck. With a head-stall, the chain is fastened to a ring over the cheek, or under the jaw, and this gives you much better control. Also, if the animal is restive, you can slip your fingers under the head-stall itself, where the straps cross and the ring is fixed, and it is very difficult for a goat to get away from this grip, as you can keep the head well down and this handicaps the animal.

A male goat should wear a head-stall from an early age and be taught to walk properly and behave itself.

And all goats, whether they wear head-stalls or collars, should be taught to walk decently. An animal that leans into its collar and pulls its owner along, or drags behind and refuses to walk, is a disgrace to its owner. Goats and dogs are very alike in this respect. They know all the same tricks,—tugging, hanging back, getting up on their hind legs, and jumping round in circles,—and the only way to make them behave decently is to take a very firm hand, grip them as closely to the face as possible, and insist on them walking at *your* pace for as long as you choose. They should, also, be taught to stand without fuss, if you have occasion to stop and speak to someone.

You must be master over your animals, or they will rule you and the results may be disastrous. How can you make an in-kid goat walk sedately, if she has never been trained to obey ? Regardless of what is good for herself in her condition, she is quite capable of racing uphill and dragging you after her, with very serious effects to herself and offspring. I have often seen this sort of thing happen and felt like throttling the silly owner ! I have, also, seen alarming things happen when a led goat got up to tricks on a public road. You must have your goats under control and bring them up to walk in an orderly manner at whatever pace you choose to set, if you are to avoid constant tussles and possible accidents.

If your goats walk well in collars and chains, with the chain hanging slackly between you, as many do, well and good. But never leave the whole length of the chain hanging. Take a twist in it with your hand fairly near the neck. If you do not do this, the most placid goat may suddenly bolt and have you over, if something unexpected alarms her. If you have no head-stalls but suddenly find yourself in need of one for a troublesome specimen, you can easily make one with a short length of rope. Make a loop to fit over the muzzle, and knot it securely. Slip this loop over the muzzle, with the knot at one side, and take the

rest of the rope round the back of the head and slip the end through the loop at the side of the muzzle, opposite the knot. If the animal is very restive, make another knot on this side. The tag end of the rope, which should not be more than about 15 ins., is for leading the animal; and the more troublesome it is, the less rope you should give her, slipping your fingers through the crossed pieces at the side of the muzzle, if necessary. Held like this, a goat can do very little in the way of misbehaving. Let her feel that your grip is firm and very determined, and use your voice to tell her to behave herself, and keep her head down if she is awkward. The piece of rope that goes round the back of the head is held in place by the eminence where Nature intended the horns to be, and it will not slip off unless you have left it too slack.

I hope all this advice about how to manage troublesome goats has not given the reader the idea that most of them are a nuisance, because this is not so. As I have said, goats are very like dogs, in this respect. Bring them up to behave in a rational manner from an early age, and you can do anything with them, and they are all the better for it. All my goats walk "like Christians", as the saying goes. But I have had some wild heathens belonging to other people under my care, and I have noticed that the worst specimens are generally owned by sentimental people who do not walk the goats themselves and refuse to let their employees discipline them, because the dear goats do not like it! Maybe they don't! Discipline is not pleasant to the wild and wilful. But a trained goat is happier than an untrained one and gives better results in every way. I hope over-sentimental owners will reflect on this, when considering the ups and downs in the milk records of their pet tuggers. A placid goat milks steadily.

The milk yield is very much bound up with the question of exercise, and many people advance this argument in favour of tethering. The more exercise that is taken, the less milk there will be. And violent exercise or sudden shock may cause the yield to drop suddenly from, say, a gallon to a mere pint or so. Tussles over walking or chasing by dogs are the sort of things that have this effect. But free range in itself, regardless of mental disturbance, will keep the yield down below what it might be ; and owners of super goats often choose the stall system with no grazing, in order to gain the highest possible records.

This system is very useful for people who want to keep a goat or two in a shed at the end of the garden but have no grazing. It is, also, the system adopted by those whose grazing is limited and who have to adopt the semi-stall method, keeping their goats in during the winter months and resting the land, as explained earlier. All that is needed besides the stalls is a little yard surrounded by chestnut fencing, or something equally strong, preferably close to the stalls. The goats walk about in this enclosure while their stalls are being cleaned and may be left there for some hours, if desired, so that they get the sun and fresh air. Food can be given them in the yards, so long as it is not put on the ground. Loose bags of wire netting are very useful for this purpose. Any small strip of reasonable size will do. Just tack it on to the outside of the paling, leaving it comfortably baggy. You put the food in from the outside, without entering the enclosure, and the goats pull it through the pales on the inside, just as they pull hay through a hayrack. Make several bags, if you run several goats in the enclosure, and attach them to the paling at intervals, otherwise there will probably be some bullying and teasing, and quiet goats may not get a fair share,

HOW TO RESTRAIN FREE GOATS

At the very other end of the scale is the free range system, where the goats are turned loose into a field and left to their own devices. Many people like this method, even though they know the goats will not milk as heavily as they would with limited exercise. The sort of people who dislike seeing rabbits in hutches and birds in cages are likely to want their goats to run free. But this is only possible when you have ample grazing at your disposal and some means of preventing your goats escaping on to other premises. Just how many goats you can run to the acre depends on the type of land, natural drainage, how long your goats spend out-of-doors, and how much supplementary food you give. It is dangerous to rely too much on figures in a case like this. As a rough guide, however, I would say not more than six to the acre. Goats, running free, foul and destroy far more than they eat, for they are definitely not careful and methodical grazers. So ample pasture is a *sine qua non,* and it must be divided into separate fields, or sections of fields, so that the herd can be shifted about whenever necessary. If they are not moved often enough, they will sicken and become unthrifty and eventually die.

Now, as regards preventing free goats from escaping, this is a difficult matter, unless you have very good fences. But there are ways in which free goats can be restrained. Probably the oldest is by the use of the hobble, a method practised with horses and camels when travellers are camping in open country. The hobble consists of two short straps with buckles, joined together by a short length of chain. The straps are fastened to the lower part of the legs, one fore and one hind. They can be used either on the right hand or the left, but they must be attached to fore and hind legs on the same side. A hobbled animal can get about to feed, but its progress is slow and it cannot clamber about walls and hedges, neither can it jump. But you have to be careful to adjust the length of the piece of chain to the animal in question, or it may be ineffective. For instance, a short-legged goat can do quite a lot of running with a long hobble.

Another way of restraining a free goat is by making it wear a " puzzle ", or " frame ". This is quite easy to make and merely consists of three pieces of wood, from two to two-and-a-half feet long, according to the size of goat. They are fastened together, in the form of a triangle, with nuts and bolts, and the ends are left projecting. Wing-nuts are best, as they are convenient to do and undo with the fingers. Only one bolt has to be undone to put the puzzle over the goat's head, and care must be taken to see it fits well, so that it does not fall off while still bolted. If it does, you need to reduce the size of the triangle, by making another hole or two. If the " puzzle " fits well, so that it cannot fall off, (or be wriggled out of, with the aid of a handy tree or fence), it is not likely the goat will break through hedges. The jutting-out ends of the triangle prevent this, while not interfering in any way with feeding. It will not, however, prevent the goat climbing or jumping.

If your trouble is that the goats stretch up after poisonous trees, you can prevent this by fixing a leather strap round the body, just behind the shoulders, letting it wear a head-stall instead of a collar, and using a third strap to link the head-stall from under the chin with the body strap, passing between the forelegs. This third strap must be carefully adjusted, so that the goat can only lift its head a little higher than the level of its body. It can then feed, but it cannot rear up and get at trees.

All these devices work well, if properly adjusted, but few people use them. They mostly argue that, if a goat must be restrained, it might just as well be tethered. Also, a tethered goat will generally milk better than one that runs its food away on free range, and tethering is certainly the thriftiest way of using pasture.

Before we leave this subject, let me give you one last word of warning. Never tether a goat with a piece of rope. Many people do so, and they may carry on for years without mishap. But any goat that is not toothless with age can bite through a rope in a few minutes, if it just occurs to her to do it!

So far, I have made no more than a passing reference to herding, and I do not propose to say much about this system. It used to be favoured by breeders who did not have their living to earn and so had leisure time to spare. Nowadays, however, there are fewer of us in this position. Even the children have their days all mapped out for them. And not many people can spare the time to go "tenting", as we say in Scotland.

Briefly, it means taking your herd out for a leisurely stroll, perhaps on their chains at first, and then letting them loose when you reach a suitable place. The goats amble about at will, browsing and playing, while the "tenter" chews a bit of grass and just watches the goats sufficiently to see they do not eat anything poisonous or get up to any mischief. It is a lazy, leisurely sort of way of passing the time,—if you have got any to spare!,—and the goats certainly have a most enjoyable outing. They get their exercise and sunshine and a great variety of food. But you must have your goats strictly under control before you practise this system. Also, you must be careful not to send them out with an attendant who is not strictly conscientious. Leaning against a fence, waiting for the time to pass, does not conduce to great alertness of mind; and, if your goat-tenter is given to dreaming, (or takes a novel in his, or her, pocket), there may be trouble.

CHAPTER SIX

FEEDING

THERE is no need to shy off keeping goats on the score of feeding, for they are about the most accommodating creatures in the world in this respect. There is precious little that comes amiss to their tastes, and many plants that are poison to other animals are good for goats. Every household has a certain amount of waste that is good goat food, and there is often waste and surplus from both flower and vegetable gardens, as well as from the orchard.

The goat's diet can be roughly divided under the following headings: —

- (i) Hay.
- (ii) Greenstuff and roots.
- (iii) Concentrates.
- (iv) Water.
- (v) Salt.
- (vi) Minerals.

Good hay is the best and safest food you can find. Many goats go all through the winter on hay and nothing else, except, of course, water and salt. Needless to say, they do not milk heavily on hay alone, but they do keep healthy and in production, for hay is a perfect food for grazing animals. All goats should " pack up " with hay before going out in the morning. Then they are less likely to come to harm if they eat wet, or even frosted, greenstuff. Hay is a fine " settler " for what follows. It is, also, the best " nightcap ", and all goats should have their racks filled last thing at night. Again, if a goat is not as well as usual or has scours (which is the stockman's name for diarrhœa in animals), the first thing to do is to knock off concentrates and succulent foods and put the patient on hay only. Hay is the safe food.

It must, however, be good hay. Goats do not turn up their noses at coarse and apparently tasteless foods, but they will refuse what is musty or dirty. If you give a goat bad hay, she will pull it down on to the ground and trample on it, as much as to say, " This is only fit for bedding ! " And, no matter how many times you may put it back in the rack, she will not eat it. She will go all day without food rather than eat what is musty or dirty.

Good meadow hay is best. Clover hay is liked by adults, but the stalks are too tough for kids, and, in any case, it is an expensive item for such untidy and wasteful feeders as goats. However much one may like them, there is no denying they are untidy feeders. They are built that way and they cannot be altered, so we just have to remember it and do the best we can to give them small opportunity for waste. When meadow hay is pulled out of the rack, it clings together fairly well. But clover hay is looser, and, as it is pulled out, showers of the valuable seed-heads fall to the ground. If they are caught in the manger in which the concentrates are fed, they will eventually be eaten ; but, too often, they fall to the ground and are trampled underfoot.

" Greenstuff and roots " covers the widest range of all and almost needs a chapter to itself. There are the obvious crops that can be grown in garden and field especially for the goats, such as cabbage, kale, mangolds, swedes, kohl-rabi, lucerne, sainfoin, clover, vetch, chicory, giant dandelion, etc.; there is a large amount of inevitable waste and surplus material from house and garden ; and, lastly, there is the vast and varied assortment of wild growth. In fact, one hardly knows where to begin. Novice goat-keepers so often write and ask me what plants in their gardens are good for goats, and I generally feel like replying that it would be easier and quicker to say which were not ! It certainly is! So, acting on that principle, I will first give the names of some of the plants that should be avoided.

Unfortunately, goats seem to have no instinct at all in this matter. They will wolf down anything green, even if it is only the flowers on your pinafore or the washing on the line! They nibble, here and there, at anything that takes their eye, even if it is a deadly poison; so it is up to you to know which plants are dangerous and see your goats have no chance of getting them. And, in this connection, it is worth making a note of the fact that certain plants, though not actually poisonous, are unwholesome at certain stages, or in certain quantities, or if taken on an empty stomach. Remember my advice about letting the goats "pack up" with hay every morning, before going out to graze or getting their greenstuff. A doubtful plant will probably settle down harmlessly on top of a good meal of hay; and many a goat has quite definitely been known to eat poisonous plants and has neither died nor been sick.

A wise general rule is to avoid all evergreens, as most of them are unwholesome and some are deadly poison. Yew is one of the worst, being a deadly narcotic poison and more dangerous when dead than green. Rhododendron is another that must be avoided. Laurel and privet are both considered dangerous, though my own goats eat freely of privet, without ill effects, at certain times of the year. I am, however, careful not to let them get at it when it is budding, in bloom, or in berry, as I consider the poisonous properties are likely to be active at these times. If in doubt about a certain plant, avoid it, especially when it is making blossom or fruit.

Exceptions among the evergreens are ivy and holly, both of which are great favourites among goats. Ivy, which is poison to so many animals, is very good for the goat and acts as a liver tonic. Holly, in spite of its sharp points, is eaten with great gusto, as are a good many other prickly specimens, such as roses, blackberries, and thistles.

To return to poisonous and unwholesome plants, laburnum, bryony, belladonna (deadly nightshade), greater celandine, hemlock, foxglove, and hellebore, must all be shunned; and, if they flourish on your land, you will do well to uproot them, year after year, until you have got rid of the lot. It is too risky to chance that no goat will ever break loose and feed on them. And though I have, myself, had goats eat one or other of many of these dangerous plants and come to no harm, I do know they can cause death; and the risk is not worth taking. Climbing, ornamental trees, so often found on the walls of country houses, are best avoided on principle. So are such doubtful things as lilac, syringa, and other ornamental bushes. Some are safe, but many are doubtful. Monkshood is definitely poisonous, and many people consider lupin as bad, though the latter is sometimes eaten with impunity.

This subject of poisonous plants is very complicated, when considered in connection with goats. One of my own goats ate a young laurel down to ground level and suffered no ill effects; yet a friend lost a favourite goat which made a run at a clump of laurels on the way home and snatched a comparatively small amount. I am convinced that a great deal depends on the stage of growth of the plant in question, since many are dangerous at one stage and safe at others. Also, parts of a plant may be bad, while others are wholesome. Consider the potato, for instance. We, and all our livestock, can eat and enjoy the tubers, but the tops are definitely poisonous.

I am, also, convinced that the poisonous properties of a plant may fail to have ill effect on a goat if it is taken in conjunction with a large and varied assortment of other items. Laburnum has carried away many a goat before a vet. could be got to her; yet I have often seen goats snatching at bits as they passed, and they have not even been sick. Doubtless, also, the age of the goat itself has something to do with what it can stand. Therefore, my advice to all goat-keepers is to be as careful as you can and try to avoid trouble but not to get in a stew if a goat snatches a bit of something doubtful. You will know soon enough if there is anything wrong.

The majority of wild plants, and many from the garden, are safe and good ; and, if you are keen on botany, you might like to make a note of the fact that the orders Compositæ, Leguminosæ, Cruciferæ, and Rosaceæ are mainly wholesome and desirable. This covers a very wide field and includes the following:—Marguerites, Michaelmas daisies, other daisies, dandelions, chicory, pyrethrum, cosmos, calendula, camomiles, helianthus, sunflowers, asters, chrysanthemum, golden rod, knapweed, cornflower, sweet peas, green peas, beans, vetches, sainfoin, lucerne, stocks, wallflowers, roses, arabis, aubrietia, hawthorn, blackberries, raspberries, loganberries, strawberries, plums, apples, and pears. You will not, of course, pull up your strawberries and cut branches off your fruit trees, merely to feed your goats, but you can remember that all prunings, trimmings, and overgrown and surplus plants of this description are good goat food.

Among the wild plants good for goats, are all the usual small plants found in the average pasture and many hedge plants, such as meadowsweet, cleavers, cow parsley, mumble, sow thistle, milk thistle, docks, dead nettles, and even stinging nettles, and thistles. Most of the deciduous trees are also good, such as oak, ash, beech, birch, hornbeam, elm, chestnut, and willow. Willow is a useful natural remedy to use, if goats have a touch of diarrhœa. Beech mast (the fruit of the beech tree), acorns, and chestnuts are all useful, but the two latter are bitter and liable to cause severe constipation, if eaten in excess. They are best collected and piled into heaps and left to ripen. Elder is much liked, but goats in milk should not be allowed to eat much of it, as it gives the milk an unpleasant flavour.

Goats are very fond of the young branches and bark of trees. If allowed to run loose where there are trees, they will quickly spoil them, by stripping off the bark as far as they can reach. It is a good idea to give them all trimmings and prunings of suitable trees and bushes, as, if they have nothing to gnaw, they will turn their attentions to their hayracks and mangers. Given a fairly large branch, a goat will eat the leaves and twiggy portions and will then bark the rest.

All bulb plants should be shunned, with the exception of the culinary onion. Goats are very fond of onions and leeks, but they should not be allowed to eat them in excess, as they have a very laxative effect. Nannies in milk are best without them altogether, though just a few will not matter. One must be careful, however, not to let milkers have too much of any strong-flavoured food, such as onions, cabbage, kale, and swedes, as they are liable to give the milk rather an unpleasant flavour. Various kinds of wild onions grow in the hedges in some parts, and you should be careful where you tether your milkers, if these plants flourish in your district.

All the vegetables we eat ourselves are good for goats, and there will be waste and surplus from both house and garden. Peelings, trimmings, cores, etc., all make good goat food, but be sure they are given in a clean condition. Goats are fastidious animals, as regards the state of their food. They do not mind rough stuff and, in fact, prefer it to the lush growth enjoyed by cows, but what they cannot endure, and will not eat, is dirty food. Doubtless this has much to do with the wholesomeness of their produce.

There are quite a lot of other odd items that goats enjoy, and probably many of them will surprise you. If you are making jam or jelly, give the goats the pulp, or the skimmings and stones from plum jam. They love crunching up hard things like this! They also love paper bags and the little cases and strips that have been round cakes. It may seem that there is not much food value in these items, but there is some, and, anyway, the goats enjoy them —so why put them in the fire? Also, a feed of paper is often craved by a goat on medicinal grounds. If she feels rather " off colour " inside, she will often eat quite a lot of brown paper or newspaper and will then feel better. The masses of paper evidently have a cleansing effect inside. All goats love stale bread and stale cake, and many enjoy banana skins and the peels of orange, lemon, and grapefruit. Also, some goats love ice creams !

Before we leave the subject of natural foods, let me give one more word of advice. Sometimes an individual goat will show plainly that she detests a certain perfectly wholesome item of food. It is not a question of faddiness but a definite dislike that never changes. For instance, I once had a goat that loathed parsnips. In such a case, do not force her. Just remember her dislike. I think, in such definite cases, the explanation is that the item disagrees with that particular goat, just as some of us cannot digest carrots.

As the natural wild growth dies down with the approach of autumn, other items come to hand that one might, on first thoughts, consider dead and useless. Take dead leaves, for instance. When they are really crisp and dry, goats love them, and they certainly contain nourishment. After all, what is hay but dried grass ? And one must remember that goats, in their natural state, live mainly on bark, twigs, dead leaves, and mosses during the winter. If you can collect plenty of clean, dry, dead leaves, put them away in sacks and hang them up from beams or rafters, so that they will keep dry for use. But beware of those evergreens !

Pea and bean haulm, whether green or dry, is enjoyed, and so are the pods and trimmings from the kitchen. But the pods and foliage of broad beans are generally not liked until they are thoroughly dry and black. All these things can be stored for winter fodder. And, when you are clearing away your tomato plants at the end of the season, both the old plants and small fruits can be given to the goats. Again, when you pull up your old cabbages and kale, and stumps from which the vegetables have been taken, this is all good goat food. But thick stalks should be split, or they may be wasted. The goats do not merely eat out the " meat ", as rabbits do. They eat the hard, nobbly stalks and enjoy them. But very thick stalks are sometimes difficult for them to manage, especially when a goat has not all her adult teeth ; so it is best to split them, to avoid waste.

CONCENTRATES

This seems to have covered the heading " Greenstuff and roots " pretty well, and we will now proceed to consider concentrates. A certain amount of this class of foodstuff is necessary, especially during the autumn and winter months, if good production is to be maintained. I do not, however, advise heavy feeding with concentrates. They should be used to supplement the amount of natural foods and hay, so that the goat has enough, beyond what is required for body upkeep, to maintain a good milk supply. Some people feed concentrates heavily—at least, they did before the war came and they were severely rationed. The daily allowance of cake and meal was worked out for each goat, according to her yield, and every effort was made to coax her to eat her full allowance, whether she wanted it or not. I do not approve of this method. Natural foods are more healthy than those which have been processed, and, though a heavy milker might have difficulty in eating enough of the bulky, natural foods alone to keep up her supply, I consider she should have as much of them as she wants. If she is forced to eat more concentrates than she fancies, she will go short on natural foods. And I think it was this carrying to excess of artificial feeding that caused the rumour to go round before the war that our goats were not as hardy as they used to be. Some were not. But a goat that lives a healthy life and has plenty of natural foods is usually as hardy as one could wish.

Concentrated foods are of two kinds —i.e., those that are rich in protein and those that are poor in protein but rich in carbo-hydrates. The protein-rich foods include the many cakes, such as groundnut or earthnut cake, linseed cake, coconut cake, cotton cake, beans, and peas, and the meals made from these substances. The starchy foods —i.e., those that consist mainly of carbo-hydrates,—include oats, wheat, barley, maize, and the meals and by-products obtained from them. The correct ratio for milk production is one part of protein to four of carbo-hydrates and this combination can be obtained by the use of any of the above meals and cakes. The reason, however, for using certain items in preference to others is that the goats like them ; and they will give better results on mixtures they like than on those they do not.

Some people save themselves the trouble of working out a ration and mixing different meals, by purchasing ready-made dairy nuts. This is a good idea, if you only have a goat or two, for the nuts are correctly balanced for milk production and you only have the one kind of food to buy and store. If, however, you are out for high records and getting the very best out of your goats, you will probably want to buy a number of things and make your own mixtures, according to the state of production of the goat in question. For instance, a very heavy milker can do with an extra feed of something she fancies, especially a little extra protein. Again, one finds that goats tire of having the same kind of food every day, even as we do, and it pays to give them a change.

If I were asked to name the goats' favourite concentrated foods, taking a fair average, I should say groundnut cake, flaked maize, crushed oats, and broad bran. On the other hand, many goats have a definite dislike for linseed cake, cotton cake, and any meals so finely ground as to be powdery. They snort into them, blowing them about in a most wasteful manner and making themselves sneeze.

Some useful formulæ for milk production are the following: —

1 part groundnut cake or meal.
2 parts flaked maize.
1 part broad bran.

—

1 part bean meal.
1 part flaked maize.
1 part crushed oats.
1 part bran.

—

1 part linseed cake.
1 part maize meal.
1 part dried brewers' grains.
1 part bran.

—

Parts are by weight, not measure. These are all useful mixtures, and they can be varied with whatever other foods are available, by substituting other meals or cakes of a similar nature. It will be noticed that I have given bran in all of them. This is not always done, but bran is a great favourite with goats and they never seem to tire of it. They seem to like anything of a flaky nature, rather than meal that is powdery, and I find that mixtures containing bran are always eaten more readily than those without. It is not essential to have broad bran, but it is liked better.

The brewers' grains mentioned are a by-product left over from brewing, and they can be bought either in the wet or dry state. Farmers buy them in the wet state for their cows, but they only keep a few days like this and goat-keepers are best advised to get the dried grains. These are obtainable from food merchants.

They are rather like grass seed to look at and have a pleasant odour. See that you do not get a musty specimen.

Never give linseed damp, for it is liable to cause prussic acid poisoning, if fed carelessly. While quite dry, it is safe ; and, when boiled (as in the case of calf meal), it is safe. But it must be well boiled, so that the poisonous properties are given off in the process of boiling. Many people shun this cake altogether, for this reason ; and, as it is not exactly popular with goats, one might just as well give them something else, if available. But remember that a good-grade calf meal, which you will probably use for weaning kids, is likely to contain a proportion of linseed meal, and see that it is always well boiled, cooled to the required temperature, and fed at once.

Corn should never be given whole, for the goat's habit is to bolt her food in a half-chewed state and bring it up later for cudding. Whole grain, taken in this way, generally causes indigestion. Oats are best given crushed, as they are given to horses. Wheat should be crushed or roughly kibbled. Barley is not a popular food, though it can be used crushed or in meal form. With both barley and oats, it is essential to see you get a good sample ; and, if you can buy it in the whole state and have it crushed or ground afterwards, so much the better. If you get a poor sample of oats or barley, it is mainly husk.

Maize, in the flaked form, is a universal favourite, resembling, as it does, the crisp corn flakes so popular on our own breakfast tables. It can, also, be used in meal form. Maize gluten feed is another variation. Some people damp the food, when meal instead of flakes is used, and this is quite a good idea, *so long as there is no linseed in the mixture*. It is more usual to give goats their food in the dry form, but some like it damped ; and, unless you have very good mangers or feeding boxes, it is certainly less wasteful to feed it slightly damped.

Sugar-beet pulp is a food that comes in a rather different category. It is the residue left after the extraction of sugar from the beet, and it has a good deal of nourishment left in it. But it is not a meal or a cake. It really takes the place of roots or greens when they are scarce, or by way of a change. The pulp is very dry and shrivelled —rather like a large edition of tea —and it is greyish or brownish in colour. It should be well soaked in hot water before use and be left, covered, until it has swollen up and is soft. Leave it plenty of time and plenty of room, for it swells up tremendously. Do not, however, make it sloppy. You will find out by experience just how much you need do for a day, how much water it requires, and how long it takes to swell. It should be left at least an hour. It can then be fed either by itself, when cool, or lightly dusted with bran or middlings.

The formulæ given earlier in this chapter are to be looked upon as useful guides. They are not meant to be hard-and-fast rules. To feed well-balanced rations should always be one's aim, because this gives better results than careless feeding. But one does not need to run to extremes and measure everything out to the last ounce. After all, what does one know of the quantity or quality of the natural foods that have been eaten on any given day? Some people think that all the fuss about balanced rations is quite a good joke —and so it is, when one considers the fuss some other people make about the balancing ! The thing to do is to strike a happy medium, knowing what the ideal balance is and endeavouring to get as near it as possible, without making a fetish of it. Remember, also, that a heavy milker is the better for a little extra protein, when in full flush.

I solved the problem of balancing rations many years ago, by getting an enamel bowl that held exactly a pound of middlings, when full. I scratched a mark to show how full it should be for a pound of maize meal, how full for fish meal, and so on. I also made a mental note of the fact that it had to be filled two-and-a-half times to make a pound of bran. I used this bowl for goat food, poultry food, and rabbit food, and I got so used to it that now I can judge a pound of pretty well anything without a measure. And this, I think, is a most desirable state of affairs. But do not let your inexperienced assistants practise it!

Before we leave the subject of concentrates, a few words on the amount required per goat should be given. Naturally, this varies a great deal with individuals, quite apart from production. Some will eat all you will give them, whether their production warrants it or not, while others do not care for artificial feeding and refuse to finish the amount allotted. It pays to study your goats and know their appetites, as well as their tastes, as one goat will rarely eat what another has left. However, as a rough guide, one may assume that an average goat, milking well, should have 1 lb. of concentrates a day for maintenance and 5 ozs. for every pint of milk given. This, of course, is in addition to hay and whatever natural foods are available. Thus, a goat giving a gallon of milk a day ought to have 1 lb. plus 40 ozs. of concentrates—i.e., 3½ lbs. daily. Whether or not you will get her to eat this amount remains to be seen. If she will not, and her milk yield continues satisfactorily, there is no need to worry. But, if it drops, or fails to rise when it should, try some other food, to coax her appetite.

MINERALS

There now remain only three other items to consider—minerals, salt, and water. Minerals are essential to health, and a goat starved of minerals will endeavour to get what she craves by scratching up and eating soil, gnawing her manger, and so on. Some people give minerals in the form of bricks. These are slipped into holders that hang on the wall above the manger, and no goat needs inviting to start licking them. Other people prefer to use a mineral mixture and sprinkle the correct allowance in with the dry food. A few even take the box around the stalls and give each goat her ration to lick out of a spoon. Solid licks are the simplest form, and you can be sure that no goat will take more than she needs and do herself harm. At the same time, if you take your goat-breeding seriously and have a number of goats of all ages and stages of development, there is much to be said for having mixtures made up to your own formulæ, so that you can vary the dose according to the needs of the animal in question.

Formulæ for mineral mixtures can always be obtained from the well-known agricultural colleges and from the offices of the County Agricultural Committees; and I do advise people to take the advice of their own county agricultural officers, since mineral needs are so bound up with the questions of soil and pasture, and your own county officers should know more about your locality than someone from another part of the country. There is one thing, however, about which I invariably warn stock-keepers of all classes. Whether you keep goats, poultry, or anything else, do not use mysterious mixtures of which you do not know the ingredients. An animal may suffer severe ill-health, and even disease, either from a deficiency or an overdose of one particular mineral. You must know with what you are dealing. If you feed strange mixtures, either in foods or minerals, you are dabbling with the unknown, unless you have the formulæ. If we buy a patent medicine for ourselves, we find the formula on the label. We should be equally strict regarding what we give to our animals.

Most mineral licks and mixtures contain a good deal of salt, and many people do not bother to give them any more. Personally, I see that every goat has her lump of rock salt to lick, in addition to her minerals, and I think this is a wise practice. An animal's needs vary, according to whether it is growing, pregnant, milking heavily, or going dry. They also vary with the time of year and the quantity, quality, and variety of food eaten, especially natural foods. The same plant is by no means of the same value at all times of the year. Again, animals that have their freedom find a certain amount of mineral substances in the earth itself. The amount thus found cannot be estimated, but the advice of your county officers and nearby agricultural colleges is bound to be helpful in this direction.

If you give each goat a mineral lick, or a certain allowance of mineral salts, and, also, see it has access to a lump of rock salt, you will find a very appreciable difference in the amount of attention paid to them by the same goat at different times. Sometimes she favours her salt and sometimes her minerals. The difference is so marked that one cannot account for it by saying it is fad or fancy. She knows what she needs, and I believe in giving her the chance of making her choice. I believe, also, in having different kinds of mineral licks or mixtures made up, so that one can give what seems best for any goat at any particular time. Some people keep a large lump of rock salt in the pasture, so that free goats can go and lick it when they like. This is quite a good idea, if your goats are untethered, as they lick at all times of the day and this encourages them to drink.

Goats are not, by nature, heavy drinkers. In fact, many goats hardly ever touch water, if they have plenty of natural foods. But, if a goat is to milk well, she must, also, drink well ; so goats must be enticed, by every possible means, to get into the habit of drinking daily. Of course, if you have reared them yourself, it is a different matter. Kids which are taught to drink water, from the time they are weaned, generally keep the habit for life. But, if you are irregular in offering water, or if you offer it in a way they do not like, they may lose the taste for it ; and a goat that does not want to drink cannot be made to do so. Then your milk yield will suffer.

Water should be offered at regular times, both morning and evening, also at midday during the hot months. Do not leave water standing about in the goat-houses, or it loses its freshness and will not be touched. Some goats like their water slightly warm; some like it salted; some prefer it with a sprinkling of oatmeal or middlings in it, while others will snort in disgust at the sight of the slightest speck! Some like rain water and some do not. There is no doubt about it that this question of getting goats to drink is often a thorny one, and it is no use pretending it is not. But it will pay you to find out how each goat likes her drink and give it her so, for that is the way to get heavy yields. Remember that milk itself is 87·2% water. But remember, also, the great value of the other 12·8% held in solution.

If you take your goats past a certain pond, stream, or even water-trough each day, it is quite likely that they will show a desire to stop there and drink. Apparently there is a fascination about something that is seen regularly. Take advantage of any such desire, and do not rush them past when they fancy a drink, just because you want to get on with the jobs. Make allowance for it beforehand. I have often got difficult goats to drink by leaving one of those big, red, bread crocks, full of clean water, at a place they passed each day. Another dodge is to have a length of guttering, to take water into a tank, or trough, and run water from the tap to the other end of the guttering with the garden hose, at the time the goats are taken out and brought in from pasture. There seems to be a great fascination about running water—and goats are not the only creatures attracted by it! But do not run the water straight into the trough itself from the hose-pipe, or one of the goats may bite the pipe in half!

CHAPTER SEVEN

BREEDING

GOATS have a definite breeding season, or rutting season, to use the correct expression. This begins about September and continues until February or March. Individual goats vary, some not having their first period of oestrum until October, or even November, and some not coming in at all after about the middle of February. September to March represents the outside limits of the normal breeding season. Occasionally one finds nannies that come on heat outside the normal breeding season —i.e., between March and September—but they are rare, and matings taking place out of season often have no effect.

When a nanny is ready for mating, she is said to be in season, in use, on heat, or ready for service. All these expressions mean the same thing, and the correct name for this short spell is the period of oestrum. It is very short in the goat, sometimes only lasting a matter of hours and never more than a day or two. So, if you have no male goat of your own, arrangements must be made in advance about getting her to a suitable billy at the right time. Fortunately, though the period is so short, it is very regular, and this simplifies the matter of making arrangements. Every three weeks, almost to the hour, your nanny will come on heat regularly, until she has been successfully mated or the season of breeding ends. So, if one mating is unsuccessful, you have a good many chances to try her again, before it is too late.

When a nanny is coming into season, she becomes restless and bleats in a peculiar manner. There is no mistaking this kind of bleat, once you have heard it. It is absolutely different from the usual bleat and has a sort of worried note. Hearing this bleat when in bed at night, I have often made a mental note of the fact that a certain goat had come into season, without even seeing her. There are, of course, other signs, but that of the voice is so infallible that I advise all goat-keepers to listen for this kind of bleat and get to know it. So much depends on getting your nanny to the male in time, and the other signs of oestrum are variable and often very slight.

Goats vary a good deal in this respect. They are restless and twitch their tails in an irritated kind of way, as though flies were bothering them. On examination, it will be found that the vulva is swollen, to a greater or lesser degree, and a slight discharge may be noticed. The milk yield may drop. Sometimes it drops suddenly and heavily; but this is nothing to worry about, as it will return to normal when the period is over. These are the signs for which you should look, if you think a goat is coming into season; but, as I say, individuals vary a good deal in these matters, and I consider the safest guide is the change in the voice. Some goats bleat a great deal at this time, while others are fairly quiet, but there are always the altered tone and the worried note.

The best time for mating is when the period is on the wane, so you need to know just how long each nanny remains in season and take her to the billy when she has just passed the peak. If you have to send her away by rail, you should make arrangements well in advance with the owner of the billy. When you write to ask about the stud fee, give the date on which you expect your nanny to be ready for service, and state that you will send her a day or two beforehand. Make this definite arrangement in advance. People who hold males at stud should have adequate accommodation for boarding visiting females for a day or so, for an animal must not be served immediately on arrival and then sent straight back. Apart from the cruelty of inflicting the return journey on the animal so soon, such a hurried mating would be likely to be abortive.

The owner of a stud male should have the necessary accommodation for females brought for service, but you ought to make arrangements with him beforehand, as otherwise you might find him unable to accept your nanny just at the time required. Having arranged the probable date all in good time, you can then send a telegram at the time of despatch, giving the departure time of the train on which you have put your nanny. You have then done all you can to avoid wasted journeys and loss of time.

If you have your own stud male, you do not have the trouble of sending your nannies away for service, but you have to be very careful indeed that neither he nor any of the females break loose and mate without your sanction. The same applies if there is any other male goat within scenting distance. Goats are incredibly strong for their size and seem to realise their powers more than ever when they desire to mate; and there are few things so infuriating as finding your beautiful pedigree nanny has effected a mésalliance with some smelly, horned, undersized little scrub of no known breed! Of course, it will get your goat into milk as well as any other billy; but, if you have a good nanny, you might just as well rear good kids from her.

To discourage the keeping of low-grade billy goats, there is the Stud Goat Scheme, sponsored by the British Goat Society and the Ministry of Agriculture and Fisheries. Under this Scheme, the services of some of the best male goats in the country are available to "cottagers, smallholders, and other persons of a similar position", at a fee not exceeding five shillings. In some cases, the fee is as little as half a crown. The actual amount is decided by the owner of the male, but it must not be more than five shillings to those who qualify under this Scheme. Those who do not can still obtain the services of the same stud males, but they have to pay the full fee, which may be anything from half a guinea to about two guineas. There is a good deal of variation in the fees charged to those who do not qualify as cottagers or smallholders, as it is then an ordinary business arrangement between two independent parties.

This Scheme is a very good one, and it is not merely a Ministerial "plan", which latter label would, unfortunately, immediately damn it in the eyes of many. It is approved and financially assisted by the Ministry; and this means that the owners of really good stud males have the financial inducement of a premium, if they will allow their males to stand at stud at the reduced fee. But the actual passing of the males, for admittance into the Scheme, is almost entirely in the hands of the British Goat Society, and the conditions and regulations are very stringent. Every stud goat accepted must be entered in the Society's Herd Book, entry into which is itself governed by strict conditions. Each male must be naturally hornless, and certain conditions regarding the purity and production of parents and grandparents must be fulfilled. The goat must be held at stud at premises approved and open to inspection by authorised officers of the Society or the Ministry at any reasonable time, and the owner of the stud male must keep a Service Book, of a type approved by the Society, in which all relevant particulars must be entered with regard to nannies served at the reduced fee under the Scheme. These books, also, are open to inspection at any time by officers of the Society and the Ministry, and premiums are only paid in respect of services given under this Scheme.

By using these males, you can be as sure as it is reasonably possible to be that you are getting some of the best blood in the country; and, whether your nannies are pedigree animals or not, it is worth getting a good male to serve them, as then the nanny kids should be worth rearing. It is possible to grade up from quite a nondescript nanny into a really good herd, by choosing first-class males every time and rearing the nanny kids.

During the years of the second world war, there was rather a lapse backwards, as regards male goats. Owing to the difficulty and danger of sending stock by rail, the shortage of petrol and feeding-stuffs, and the absence of many breeders, through the call-up into the Forces and for war work, quite a crop of non-pedigree billy goats appeared. It was so much easier to walk a nanny to a convenient billy than to try and get her to one on the other side of the county, and the main thing was to get the nannies into milk. Few people wanted to rear kids in wartime. So the half-pedigree, and even the scrub, had nannies coming to the door, so to speak, and there is no denying that some of these males were fine, handsome fellows. In a way, it would have been strange if they had not been a good deal better than the smelly billies seen in many a country lane and field, during the days of the first world war. The great improvement that took place in goats between the two wars was due, in no small measure, to this Stud Goat Scheme. It had run for 18 years, when it had to be temporarily suspended in 1939, and during, those years, it must have supplied good blood to many thousands of non-pedigree nannies which would otherwise have been mated to the nearest scrub. In recent years, about 2,000 nannies per annum have been served under this Scheme.

Undoubtedly, the non-pedigree males that raked in the shekels during the second world war were, taken as a whole, a much better lot than their predecessors of the early part of the century. But their time has gone now; and every goat-keeper, whether large or small, ought to get good service for his nannies. Stud males are held under this Scheme in most counties in England and in some parts of Wales. In some counties, there are a number of different males at stud under the Scheme. There are probably several within convenient distance of you, and this gives you a choice of breeds and strains.

There is just one other point to be mentioned, with

regard to sending nannies under this Scheme. It is stated in the Conditions that "There is no obligation on the part of the stud goat owner to board or feed visiting female goats, upon payment or otherwise, and no female goat shall be left at the premises of the stud goat owner other than by mutual agreement and on the owner of the female goat providing food for same". This clause has caused a good deal of heart-burning and has, to my knowledge, resulted in many people shunning the Scheme and using less desirable billies.

It certainly would be a callous and cruel thing to take a nanny on a journey, mate her at once, and take her back immediately on the return journey. Further, many nannies will not accept service when tired or upset, as they often are after an unusual experience like a journey, either by road or rail. Finally, to take an animal on a journey immediately after service often renders the service ineffective. Viewed from this angle, the clause appears to have been conceived without any consideration for the poor nanny. Actually, it was inserted to protect the stud goat owner from financial loss, due to the thoughtlessness or carelessness of many owners of nannies. Believe it or not, there are quite a lot of people who will take or send a nanny some days before she is due for service, claim the service at the reduced fee, and then calmly expect the owner of the stud male to feed and board the nanny out of the five shillings—or half a crown! Some even quibble about paying the return fare for the nanny!

This is absolutely out of all reason. It should be clearly understood that the owner of a nanny sent for service pays the fare of the nanny *both ways.* And, if you want to have your nanny boarded and fed for a day or so—which is only reasonable—you must make such arrangements, in writing, with the owner of the stud goat before you send your nanny. Further, if you are claiming service at the reduced fee under the Scheme, you must expect to pay something for your nanny's keep.

It is different if you pay the full fee. People who hold males at stud,—whether they are goats, or dogs, or anything else —have a duty to visiting females, and one is to see that they are well cared for and properly fed. And people with a reputation to maintain generally do their duty in this respect, for it is not to their advantage to let the service given by the male be nullified by callous treatment of the fee-bringing visitor. One's *normal* fee is, therefore, fixed to cover a certain amount of expense in this direction, and it is only when a female is left beyond a certain limited time that one charges for keep.

As regards sending food with the nanny, I have never yet known of this being done. It is simpler to feed the visiting nannies as one feeds one's own goats, than to have odds and ends of this and that rolling up with each visitor. However, the point to remember is that you must not expect board and lodging thrown in for luck, when you only pay five shillings, or less, for service; and it is up to you to make your own arrangements, in writing, beforehand. If a stud goat owner sounds awkward and disobliging, try another one! There are plenty of them!

Just how soon a goatling can be mated depends, to a certain extent, on the time of year at which it was born; but, generally speaking, they should not be served until they are fifteen months old. They begin to have the usual periods of oestrum much earlier than this, sometimes when as young as four months, but they must not be allowed to mate at such an early age. If accidentally mated when very young, they may not survive the kidding. If about a year old when they have their first kids, they will probably pull through safely; but the kids will be small and weak, and the mothers generally remain undersized for life and seldom make good milkers.

As kids and goatlings, they should be allowed ample time for growth and development, so that they make good, roomy bodies, with plenty of space for dealing with large quantities of food. Goats could never be fed on what is often facetiously called "the tabloid system". Much of the food they eat has comparatively little food value; but they need the bulk, and they must have large frames and the capacity for dealing with it, if they are to become heavy milkers. So never mate a goatling, no matter what her age, if she is not already a good size. Let her have more time for growing, and mate her later in the season.

There are, however, occasions when a goatling may well be mated before she is quite fifteen months of age. If she was born round about the end of January, and if she is well grown and well developed, she could be mated at the beginning of March—i.e., when she is about thirteen-and-a-half months of age. If this is not done, you will have to keep her another six months before you get the chance of mating her, which means she will be over two years old before she comes into profit. So, if you have your nannies mated early in the season, do all you can to push on the resultant nanny kids and get them a good size, so that you do not have to run them on into a second breeding season before getting them in kid.

This sort of thing needs bearing in mind when planning your breeding programme. If you just mate in a haphazard manner, without any definite plan, you may find yourself with flush periods, when you have more milk than you know what to do with, interspersed with spells when you have none, or very little. This is bad management. If you have not less than three goats, you never ought to be without milk, barring unexpected bad luck. A good breeding programme means getting your young goatlings well grown, getting them stocked (i.e., in kid) without unnecessary delay, and having a regular and unfailing supply of milk.

If you only want milk for home use, two goats may be sufficient; but, if you like to make a certain amount of butter and cheese as well, you ought to have

BRITISH GOATLING
Didgemere Dhora. The property of Mrs. Arthur Abbey.
[*" Sport and General " photograph.*

ANGLO-NUBIAN GOAT
Theydon Butterfly, Q*. The property of Miss K. Pelly.

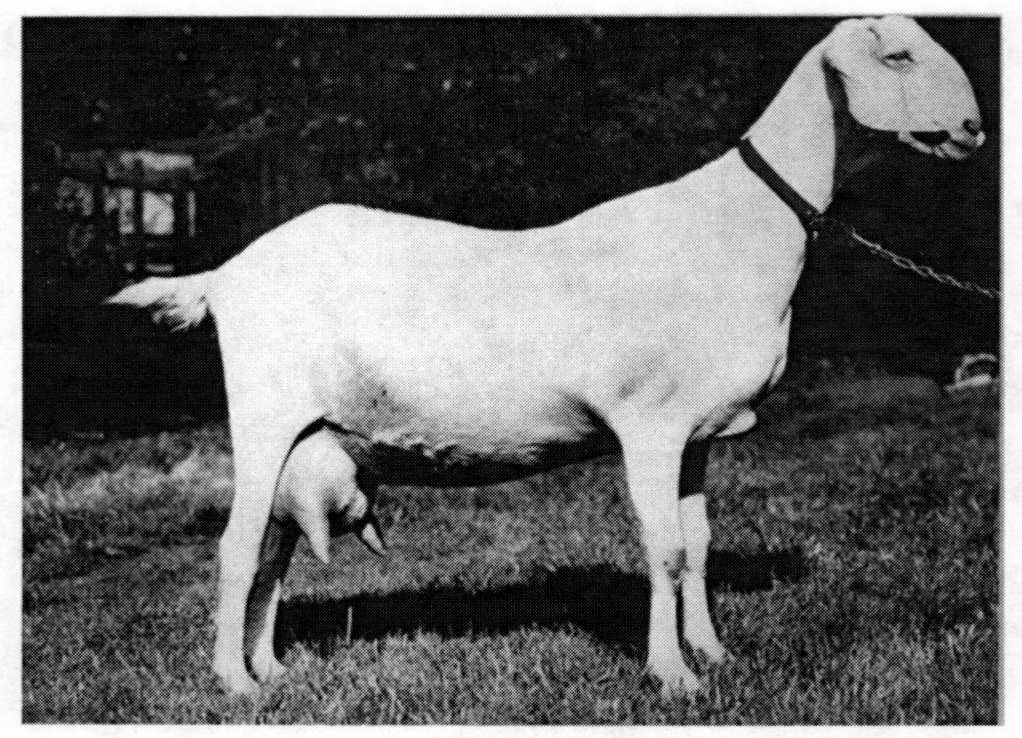

ANGLO-NUBIAN GOAT

MR2 Theydon Bellaritza, Q*Q*Q*Q*Q*. The property of Miss P. Kelly.

not less than three. They should then be mated in a definite order. Goat A should be mated in September, at the very beginning of the breeding season, and she will then kid in February; Goat B should be mated in February, (or January, if preferred), and she will kid in June or July; Goat C should be left unmated and carry on till the following breeding season. Needless to say, Goat C should be the one that is milking best.

By the time the next breeding season begins, Goat C will probably be either dry or nearly so, and she should be mated in September; Goat A should be mated in January or February; and Goat B should be left to milk through, unmated. In the third season, Goat B is mated in September, Goat C in January or February, and Goat A runs on unmated. You continue this sequence, year by year, each time letting one goat run on unmated. As this goat is the one that was mated the previous January or February, she certainly ought to be able to carry on and keep milking through the winter; and then her yield will naturally rise with the coming of the fresh greenstuff. No matter when a goat kids, her yield will tend to drop in the winter and pick up again in the spring.

The reason alternative months are suggested for the mating of the second goat is that it is risky to leave a goat unmated till the end of February, if you want to be sure of getting her in kid. Some goats continue to come into season during March, but some do not; so there is no certainty about the state of those mated in February. If a goat is mated and conception does not take place, she should come into season again three weeks later. You then know she is not in kid. But it sometimes happens that a mated goat goes six weeks, before coming into season to show she is not in kid. So it only needs a little simple arithmetic to show that a goat mated in February, or even late January, cannot be presumed to be in kid because she did not come into season after the service. For this reason, it is wise to have the nannies marked down for late service mated during the first half of January. When a goat comes into season again after having been served, it is said that she turns. This is the usual expression among goat-keepers and merely means that she returns into use and must be mated again.

Goats vary a good deal in the length of their lactation. This is mainly hereditary, and length of lactation,—i.e. of the period for which a goat remains in milk,—is one of the points breeders are always seeking to improve. Nowadays, a good goat will remain in milk for about two years after kidding, and some cases have been known of them keeping on for three years. Goats that need mating every year are a nuisance and not worth keeping. Of course, feeding influences length of lactation, and so does the time of year at which kidding takes place, and another factor to be taken into consideration is the length of rest the goat had before kidding. Animals bred from high-producing stock are often very hard to dry off, prior to kidding. Sometimes one simply cannot dry them off, except by cutting down the food supply to such an extent that the goat loses condition, and this affects the next lactation. With such persistent milkers, it seems best to let them go on, if they cannot be stopped by less drastic means ; but a break of six weeks before kidding is desirable.

People with little experience of goats are often surprised to find that there is such a thing as a maiden milker. This is a young goatling that comes into milk though she has never been mated, and they are by no means unusual among stock bred for high production. If young animals come into milk like this, they should be milked in the ordinary way, as it is dangerous to leave them with full udders. Apart from anything else, it is liable to start the troublesome habit of self-sucking, which is often very difficult to cure. Milk them, if you see they need it, but get them mated as soon as possible. Actually,—strange as it may seem,—

there has even been a duly authenticated case of a stud male giving milk !

If a goat does not come into season regularly, there is probably something wrong with her, and she should be examined by a veterinary surgeon. The average goat is very prompt and regular, but there are cases of animals showing no desire to breed, and these can often be easily and successfully treated, if taken in time. If your nannies do come on heat but are late in beginning, it may be that the feeding is inadequate, that there is a mineral deficiency, or that the animals are anæmic, possibly on account of lice. If you suspect anæmia, turn up the lips and look at the gums and inner sides of the lips. In a healthy goat, they should be nice and pink.

If your nannies seem in perfect health and yet keep on " turning ", it may, of course, be the male that is at fault. Billies can be used for stud as early as six months of age, for a limited number of females, but many billies are not stock-getters. Why this should so often happen in the goat world it is difficult to say, but the fact remains that many males prove useless in this respect. So, if you have a young billy you want to keep for stud work, it is best to try him out as early as possible. If he is not a stock-getter, there is nothing for it but to let him go for meat. A kid under six months of age is good for the home larder ; but, once they get to six months, the meat is liable to be strong and unpleasant in flavour and is only suitable for dogs. Some people shoot them themselves, skin them, and cut them up. The skins make lovely rugs, and the meat comes in very handy if you breed dogs or cats. But, if you feel you cannot tackle the job, or if you have no use for the meat, the only thing to do is to sell the unwanted billies alive to knackers. They buy old cows, and horses, and any meat unsuitable for human consumption, and their advertisements are usually found in the county papers.

CHAPTER EIGHT

THE IN-KID NANNY

PREGNANCY is not a disease, and the in-kid nanny should not be cossetted and coddled. In fact, the more natural the life she leads while she is in kid, the less likelihood there is of complications arising. Goats seldom have any trouble producing their young ones, and many a nanny drops her kids, in stall or field, without anyone knowing anything about it. When things go wrong, it is often because the expectant mother has been allowed to lie about her stall, instead of exercising in the fresh air.

As was explained in the previous chapter, you cannot be sure that a nanny is in kid, unless she has quite definitely missed two periods when she should have been in season. If she was mated so late that there is some doubt about this, you can only hope for the best and watch for other signs ; and these other signs are often very indefinite. A goat is, at the best of times, rather angular in shape, and the alteration in her figure is so slight that it is risky to take this as a guide. Many an expert goat-keeper has been unable to say, right up to the very kidding date, whether or not a certain nanny was in kid. Often it really is impossible to tell, short of an internal examination.

There are some things which may serve as guides, but I should hesitate to label anything but an internal examination a definite proof, and internal examinations ought not to be performed by unqualified persons. If a goat is in kid, it is likely that her milk yield will begin to drop about halfway through pregnancy and continue steadily to diminish. She may, also, take a dislike to walking and want to lie about and eat her food from ground level. This must not be allowed, for a goat that takes insufficient exercise, when in kid, is liable to get so soft that her muscles will not respond, when called upon for the expulsion of the kids. This means a long and difficult labour, perhaps even involving the use of instruments.

Never force a goat which you believe to be in kid to take long walks, and do not hurry her. But, on the other hand, see she takes a reasonable amount of exercise, and do not let her have her food at ground level. Put it where she must stand up to eat it.

The appetite of an in-kid goat does not necessarily increase. This depends, to a certain extent, on whether she is keeping up her milk yield. If her yield drops and her appetite is maintained, or increases, it certainly looks as though she is in kid. But, with highly bred goats, the yield does not always drop. Sometimes it even improves a little after mating, dropping gradually during the second half of pregnancy but not ceasing altogether. This brings you face to face with a very difficult problem, if you are pretty sure the goat is in kid. She ought to have six weeks' rest before kidding, and the usual way to dry off a goat is to milk her at irregular intervals and leave the strippings. For instance, you might milk her in the morning but not in the evening for a few days, watching all the time to see she is not troubled by the retention of the milk. If she is all right, you then miss a day altogether.

Treated like this, the goat should gradually dry off, so that, after a few weeks, she is quite dry and can be left unmilked for the six weeks before she kids. But, if you have a goat that will not dry off, even though you milk her irregularly and reduce her ration of concentrates, what can you do ? You can knock off her concentrates altogether ; but this may cause her to get into such poor condition that she has a bad time kidding. Or, at best, even if she gets through her kidding all right, she will be some time regaining her former condition ; and, in the meantime, you will be losing milk, and the lactation, as a whole, will probably not be as good as her previous one. If you cannot get a goat to dry off by the usual methods,—i.e., just by milking irregularly, leaving the strippings, and reducing the concentrates,—it is best to let her carry on, relieving her of milk whenever necessary and feeding her well enough to keep her in condition.

If the yield drops naturally during the course of pregnancy, this trouble does not arise; and, as she dries off, you will probably find that her taste for concentrates loses its keenness. She may, also, be harder to tempt with water, especially if she has ample greenstuff. If, however, the goat will drink really cold water, you may find out the answer to your question; for, as the mother drinks cold water, the kids may be seen to move. This does not always happen, but it does sometimes, and that really is a sign.

Do not let in-kid goats have sudden frights or unpleasant disturbances. Bustling by the attendant or badgering by dogs is liable to result in a miscarriage. And do not give worm medicines at this time. Anything of this sort should be done beforehand, if necessary. Hoof trimming is another thing that should be seen to in advance. A quiet, peaceful life, with a reasonable amount of steady exercise, is what the in-kid goat should have, and any sudden exertion or excitement should be avoided.

The period of gestation for goats is one-hundred-and-fifty-one days, but the kids may arrive any time after the one-hundred-and-forty-fifth day, so preparations should be made in good time. A first kidder may only have one kid, but twos and threes are more usual at subsequent kiddings. Fours are by no means uncommon, and fives and sixes are not unknown. However, these large families are not desirable, as the kids are necessarily very small and are often weak. Decide in advance what you will do with the kids and have a box ready for them. If you are going to take them straight away from the mother, it is best that she should not see them and that they should be taken away out of sight and hearing.

Actually, there is little extra to be done as kidding time draws near and often little to be done at the time of parturition. Have the goat well groomed and see that her udder is clean. If it is very tight and shiny, it may be necessary to draw off a little milk, but do not, on any account, milk her dry. Just take enough to make her comfortable. Let her be in a place by herself, untethered, and give her plenty of clean bedding. She should be in the place where she is to kid some days in advance.

As kidding time draws near, you will naturally be wondering when the event will take place and what signs there will be to guide you. If your nanny has shown no " figure ", you may even still be wondering if there are going to be any kids at all! However, hope for the best and get ready, and keep a watchful eye on her. Individuals vary a good deal, but probably one or other of certain symptoms will be noticed. If the goat has shown a " figure ", this appears to decrease a few days before kidding. This is because the kids have dropped into the position from which they will be born, and so the flanks of the mother have a hollow appearance. Another sign is a discharge from the vagina, which may begin a few days or a few hours before parturition. This, however, is often so slight that it is not noticed.

The filling of the udder is a sure sign that things are getting close, but, even so, this may happen either a few days or a few hours beforehand. I think the surest sign of the imminence of labour is the frantic packing up with food that goats indulge in when they know their time is very near. They eat in a way they do at no other time, cramming down hay and anything else they can find, as though they expect to be several days without food. Yet they do no fasting and often help themselves from their hayracks between the birth of the kids. However, they seem to feel that the busy time ahead may leave no time for eating, and I take this as a sure sign that the goat will shortly be in labour.

Soon after this, the goat will begin to get restless. If she is a first kidder, or if she is of a fussy disposition or has been treated as a family pet, she may keep bleating for someone to stay with her. Other goats prefer to be alone. Actually, there is nothing one can do at this stage, and, though some goat-keepers like to stay with their nannies all the time, I, personally, think they are best left alone. One should look in, from time to time, to reassure them and to see that all is well, but I think the expectant mother, (whether caprine or canine), is more often excited than soothed by the constant presence of a human being. They fuss for one to do something, though there is nothing one can do ; whereas, if left alone and looked at from time to time, they seem much quieter.

The nanny will now get very restless, nosing and pawing her bedding, lying down, and then getting up again in a few minutes, as though she cannot find comfort anywhere,—which, indeed, she cannot, for she is now in the first stage of labour. There is still nothing one can do for her, except keep an eye on her, to see that things go on all right. There should now be a noticeable discharge from the vagina, and soon the second stage of labour will begin and the goat will start straining. Although this is more severe, the average goat ceases fussing at this stage and stands steadily and concentrates on the business in hand. She can feel the kid moving along and knows she is now getting somewhere. Do not worry her with attentions, but just remain handy to see that all is well.

In most cases, the actual birth is surprisingly sudden. As the mother is straining, the bag, which is made of transparent membrane, appears, and the muzzle of the kid can be seen within. The goat generally rests a few minutes, at this stage ; she then gives another strain, and the bag bursts and the kid falls to the ground with a sudden plop. Sometimes, the bag breaks inside the goat, and the head, when it appears, is plainly seen to be lying on the forefeet, which is the correct position. In either case, the actual birth is usually quick and sudden; and, almost before you know it, there is a wet and wriggling mass of legs lying on the floor. It is surprising that the kids come to no harm this way; but they do not, and it is usual for the mother to stand at the moment of delivery.

If the mother is free, as she should be, she will probably turn round at once and begin to lick the kid. Sometimes, however, they ignore them. In any case, it is best to be with the goat at this time, and the kid should be picked up immediately and cleaned. You can make a quicker job of it than the mother ; and the second kid often comes along very quickly, in which case, if the first kid is not finished, the mother leaves it where it lies, while she begins straining again.

Some pieces of clean sacking and a rough towel should be to hand, and the mouth and eyes of the kid should be cleaned first. Then do the whole body, using the sacking to remove the worst of the thick mucus that covers it and finishing off with a brisk rub with the towel. Be very careful not to pull the navel cord. Some people use newspaper first, instead of sacking. When the kid is pretty dry and the coat looks fluffed up, put it in a hay-lined box, which should be ready and waiting. Place the box where the mother can see it, and, if she shows a desire to lick the kid, let her do so. But do not leave her alone with it in a place where she can get at it, or she may head the box over and ill-treat the kid. This is by no means unusual. If you mean to leave the kids with the mother for a while, it is a good idea to partition off a small section of the shed with a couple of hurdles and put the box containing the kids in this enclosure. The mother can then see them but cannot interfere with them.

Of course, if you are intending to rear the kids by hand from birth, or to dispose of them in any other way, it is best to take them straightaway, as soon as they are born, and not let the mother see them, as she is then less likely to fret. In this case, the procedure is the same, as regards cleaning them, and rubbing them dry, and putting them in a hay-lined box, but it should be done elsewhere. Wrap the new-born kid in one of your pieces of clean, dry sacking and carry it away to wherever you plan to keep them, which should be out of sight and sound of the mother. Have your cloths and the hay-lined box ready to hand, and proceed as already explained.

Do not have too large a box for the kids. Remember they have just come from a snug, warm nest, where they were all curled up together, and they do not like a lot of chilly air round them. Novices are generally amazed at the way three or four kids will curl up into a small box, the size of the average shopping basket, and prefer it to anything bigger. The smaller the box, within reason, the better for the kids during the first week or so. But they grow rapidly and will soon show when they want more roomy sleeping quarters.

As each kid comes along, rub it down, as described, and put it in the box with its fellows, where it will quickly go to sleep. That is all they want for some hours. In the meantime, give your attention to the mother. When it appears that there are no more kids to be born, make her an oatmeal drink. It is made as follows:—Take a double handful of oatmeal, (preferably medium), and scald it with boiling water. Leave it to stand for a few minutes and then add cold water, stirring all the time, until it is blood heat. The nanny will be thankful to have this and will probably drink the lot and lick out the oatmeal at the bottom. Give her a rack of good hay and leave her eating this while you make her a bran mash. This is made by scalding about a couple of pounds or so of bran with boiling water, covering it, and leaving it to swell for about ten minutes. Then dry it off with middlings, making it crumbly but still damp, and give it to the nanny while comfortably warm. This, also, will be joyfully welcomed, and it has a slightly laxative effect, as well as being soothing and comforting.

Give some clean bedding and leave her to rest for a while,—for the kids do not need anything in the way of nourishment yet,—and watch your nanny to see that she passes the afterbirth. It should come fairly soon, and most goat-keepers clear it away as soon as they see it. If this is not done, the goat will devour it. Actually, there is no reason why she should not. It is Nature's arrangement for righting the system after parturition, and it appears to be common throughout the animal kingdom. I have never known an animal come to harm through eating the afterbirth, but I must admit I remove them myself, when I see them, as it is really revolting seeing a goat eat such stuff. However, it is up to each goat-keeper to decide for himself what to do about this. The only things that really matter are to see that the afterbirth has been passed and to watch your goat and see she does not get constipated after kidding. Eating the afterbirth prevents the latter trouble.

With the passing of the afterbirth, kidding is over, and some more clean bedding should be given. But do not disturb the nanny by attempting to clean out the whole place. Just take away any soiled portions and replace with fresh stuff. Some people wash the goat's hinderparts with warm water and a mild disinfectant, but some do not. If much soiled, it is best to do so, if the goat will let you; but some object very much, and, if the soiling is only slight and you are single-handed, it is better to leave her in peace, rather than upset her by forcible handling.

Should the afterbirth not be passed, something must be done about it, or you may lose your goat. Sometimes it comes very soon after the birth of the last kid, and sometimes not for an hour or two. If more than four hours have passed and still the afterbirth has not appeared, you should either get in touch with your veterinary surgeon or, if you are sufficiently experienced, insert a cleansing pessary or give a vaginal douche. This should be done three times a day, until the afterbirth, (or cleansings, as they are sometimes called), is brought away. If you have never used either pessaries or enema and have nobody to show you how it is done, you should have your veterinary surgeon to attend the case and give you the necessary instructions. It is too risky tinkering about with the inside of an animal, with nothing but the printed word to guide you. This is one of those cases where an ounce of practice is worth a pound of theory.

The same applies to difficult kiddings, about which I do not propose to say much, for the same reason. If a goat cannot produce her young in the normal manner, it is often because they are in the wrong position for birth. If this is the case, the mother may go on straining until she is exhausted, but she cannot bring forth the kids. Something must be done for her. The normal presentation is with the head laid on the two forefeet. Sometimes the forefeet are found to be over the head, and these must be brought down into the correct position before the kid can be born. Sometimes the head is twisted backwards. Sometimes the limbs of different kids are entangled. These are serious cases and need expert attention and perhaps, also, the use of instruments, which, in my opinion, ought never to be used by an unqualified person. If your goat is in real trouble, admit yourself beaten,—it is no shame,—and get a veterinary surgeon to her as quickly as possible.

If you are in an out-of-the-way place and cannot get hold of a vet., the next best thing is a capable and experienced shepherd. These men spend their lives among their sheep, which are very similar in conformation to goats but, unlike goats, very often in need of assistance from the shepherd when they have their lambs. The assistance required is not necessarily a serious matter, involving the use of instruments. It may be only a case of lifting out the front feet, which are sometimes bent under, instead of being laid out flat with the head resting upon them, as they should be. Actually, this is a simple matter, if you know what you are doing, but no inexperienced person ought to tinker about with the inside of an animal, in the hope that things will come right. If you want to be able to help your goats in this way, you should have practical, first-hand instruction from somebody who can *show* you how it is done.

If there is any misplacement that is hindering birth, the kid has to be pushed back into its mother first, before any investigation can be made or the tangle put right. There is no room in the vagina to do this. When the matter has been righted, the goat will start to strain again and should eventually bring the kid to the light of day. But, as she has to go through the whole business a second time, it is very necessary not to delay in getting help, if you see something is hindering her in the first place. Do not let your goat get exhausted before you seek aid, or you may lose both her and the kids.

If a goat is obviously finding it hard going and you know things have now been put right,—or if she is in poor condition, or weak in the muscles through lack of exercise,—you can assist her by leaning over her back, facing her rear, putting your arms round her body, on either side, and clasping firmly each time she strains. And, once the feet of the kid appear, you can take a grip of them and draw gently each time she strains. This prevents the kid slipping back and so giving a tired mother the job of bringing it along again. But do not pull hard, and only draw it when the mother is actually straining. You will need a cloth to put round the feet, or you will find it difficult to keep a grip.

After an assisted kidding, the nanny will need to have an enema twice daily, for about a week. Warm water is used, about a pint at a time, with a mild disinfectant, such as is used for human beings. On no account use coarse, strong disinfectants, meant for cleaning purposes. The enema required is the kind made for human use, with the vaginal extension.

Before we leave the subject of kidding, a few words on false conception will not be out of place. For some reason or other, this is by no means uncommon among goats, and it is extremely disappointing, especially to the small goat-keeper with only one or two goats. The nanny comes into season and is mated, and, as she does not come into season again, it is presumed she is in kid. She develops a " figure ",—often an exceptionally large one,—and everything seems to be going on well. But the time for kidding arrives and there are no kids, neither does the figure go down at all.

Actually, the uterus has filled with fluid, and, until this has been passed out, it is impossible for the goat to conceive. She will probably go on past the date on which she should have kidded, perhaps as much as a month longer ; and then, one day, you will come down and find her flat and normal,—but with nothing to show for it ! It is most irritating ! The water has been passed out, and you have no kids, and not even any milk, as you will have dried off the goat, preparatory to kidding.

She will probably come into season a few days after getting rid of the fluid, but, if she has any discharge, she should not be mated until three weeks later, when it is likely she will conceive. Even maiden goatlings which have not been served are liable to develop false conception, and the owner generally supposes she has somehow been served without his knowledge. Short of an internal examination, there seems no way of distinguishing between the false and the real thing, though one's suspicions may be aroused if a mated nanny fails to dry off, in spite of an obvious " figure ". However, since it is often difficult enough, at any time, to dry off a really heavy milker, this is only a vague and uncertain guide. Generally, there is no doubt at all in the mind of the goat-keeper ; and it is not until kidding date is well past, and still no young have been produced, that he realises that it is he, and not the nanny, who is " holding the baby " ! In other words, she has just " led him up the garden " !

OBSTETRIC TABLE

CHART SHOWING APPROXIMATE KIDDING DATE FROM ANY GIVEN DATE OF SERVICE.

Date of Service.		*Kidding Date.*		*Date of Service.*		*Kidding Date.*	
Jan.	1st.	June	1st.	Feb.	1st.	July	2nd
,,	2nd.	,,	2nd.	,,	2nd.	,,	3rd.
,,	3rd.	,,	3rd.	,,	3rd.	,,	4th.
,,	4th.	,,	4th.	,,	4th.	,,	5th.
,,	5th.	,,	5th.	,,	5th.	,,	6th.
,,	6th.	,,	6th.	,,	6th.	,,	7th.
,,	7th.	,,	7th.	,,	7th.	,,	8th.
,,	8th.	,,	8th.	,,	8th.	,,	9th.
,,	9th.	,,	9th.	,,	9th.	,,	10th.
,,	10th.	,,	10th.	,,	10th.	,,	11th.
,,	11th.	,,	11th.	,,	11th.	,,	12th.
,,	12th.	,,	12th.	,,	12th.	,,	13th.
,,	13th.	,,	13th.	,,	13th.	,,	14th.
,,	14th.	,,	14th.	,,	14th.	,,	15th.
,,	15th.	,,	15th.	,,	15th.	,,	16th.
,,	16th.	,,	16th.	,,	16th.	,,	17th.
,,	17th.	,,	17th.	,,	17th.	,,	18th.
,,	18th.	,,	18th.	,,	18th.	,,	19th.
,,	19th.	,,	19th.	,,	19th.	,,	20th.
,,	20th.	,,	20th.	,,	20th.	,,	21st.
,,	21st.	,,	21st.	,,	21st.	,,	22nd.
,,	22nd.	,,	22nd.	,,	22nd.	,,	23rd.
,,	23rd.	,,	23rd.	,,	23rd.	,,	24th.
,,	24th.	,,	24th.	,,	24th.	,,	25th.
,,	25th.	,,	25th.	,,	25th.	,,	26th.
,,	26th.	,,	26th.	,,	26th.	,,	27th.
,,	27th.	,,	27th.	,,	27th.	,,	28th.
,,	28th.	,,	28th.	,,	28th.	,,	29th.
,,	29th.	,,	29th.	March	1st.	,,	30th.
,,	30th.	,,	30th.	,,	2nd.	,,	31st.
,,	31st.	July	1st.	,,	3rd.	Aug.	1st.

Date of Service.		*Kidding Date.*		*Date of Service.*		*Kidding Date.*	
March	4th.	Aug.	2nd.	April	14th.	Sept.	12th.
„	5th.	„	3rd.	„	15th.	„	13th.
„	6th.	„	4th.	„	16th.	„	14th.
„	7th.	„	5th.	„	17th.	„	15th.
„	8th.	„	6th.	„	18th.	„	16th.
„	9th.	„	7th.	„	19th.	„	17th.
„	10th.	„	8th.	„	20th.	„	18th.
„	11th.	„	9th.	„	21st.	„	19th.
„	12th.	„	10th.	„	22nd.	„	20th.
„	13th.	„	11th.	„	23rd.	„	21st.
„	14th.	„	12th.	„	24th.	„	22nd.
„	15th.	„	13th.	„	25th.	„	23rd.
„	16th.	„	14th.	„	26th.	„	24th.
„	17th.	„	15th.	„	27th.	„	25th.
„	18th.	„	16th.	„	28th.	„	26th.
„	19th.	„	17th.	„	29th.	„	27th.
„	20th.	„	18th.	„	30th.	„	28th.
„	21st.	„	19th.	May	1st.	„	29th.
„	22nd.	„	20th.	„	2nd.	„	30th.
„	23rd.	„	21st.	„	3rd.	Oct.	1st.
„	24th.	„	22nd.	„	4th.	„	2nd.
„	25th.	„	23rd.	„	5th.	„	3rd.
„	26th.	„	24th.	„	6th.	„	4th.
„	27th.	„	25th.	„	7th.	„	5th.
„	28th.	„	26th.	„	8th.	„	6th.
„	29th.	„	27th.	„	9th.	„	7th.
„	30th.	„	28th.	„	10th.	„	8th.
„	31st.	„	29th.	„	11th.	„	9th.
April	1st.	„	30th.	„	12th.	„	10th.
„	2nd.	„	31st.	„	13th.	„	11th.
„	3rd.	Sept.	1st.	„	14th.	„	12th.
„	4th.	„	2nd.	„	15th.	„	13th.
„	5th.	„	3rd.	„	16th.	„	14th.
„	6th.	„	4th.	„	17th.	„	15th.
„	7th.	„	5th.	„	18th.	„	16th.
„	8th.	„	6th.	„	19th.	„	17th.
„	9th.	„	7th.	„	20th.	„	18th.
„	10th.	„	8th.	„	21st.	„	19th.
„	11th.	„	9th.	„	22nd.	„	20th.
„	12th.	„	10th.	„	23rd.	„	21st.
„	13th.	„	11th.	„	24th.	„	22nd.

NEUTER GOAT,
TRAINED TO
HARNESS
The property of T. A.
Urie.

[" *Stockport Advertiser* "
picture.

FEEDING THE HERD
The property of Ernest S. Smith.

Date of Service.		*Kidding Date.*		*Date of Service.*		*Kidding Date.*	
May	25th.	Oct.	23rd.	July	5th.	Dec.	3rd.
„	26th.	„	24th.	„	6th.	„	4th.
„	27th.	„	25th.	„	7th.	„	5th.
„	28th.	„	26th.	„	8th.	„	6th.
„	29th.	„	27th.	„	9th.	„	7th.
„	30th.	„	28th.	„	10th.	„	8th.
„	31st.	„	29th.	„	11th.	„	9th.
June	1st.	„	30th.	„	12th.	„	10th.
„	2nd.	„	31st.	„	13th.	„	11th.
„	3rd.	Nov.	1st.	„	14th.	„	12th.
„	4th.	„	2nd.	„	15th.	„	13th.
„	5th.	„	3rd.	„	16th.	„	14th.
„	6th.	„	4th.	„	17th.	„	15th.
„	7th.	„	5th.	„	18th.	„	16th.
„	8th.	„	6th.	„	19th.	„	17th.
„	9th.	„	7th.	„	20th.	„	18th.
„	10th.	„	8th.	„	21st.	„	19th.
„	11th.	„	9th.	„	22nd.	„	20th.
„	12th.	„	10th.	„	23rd.	„	21st.
„	13th.	„	11th.	„	24th.	„	22nd.
„	14th.	„	12th.	„	25th.	„	23rd.
„	15th.	„	13th.	„	26th.	„	24th.
„	16th.	„	14th.	„	27th.	„	25th.
„	17th.	„	15th.	„	28th.	„	26th.
„	18th.	„	16th.	„	29th.	„	27th.
„	19th.	„	17th.	„	30th.	„	28th.
„	20th.	„	18th.	„	31st.	„	29th.
„	21st.	„	19th.	Aug.	1st.	„	30th.
„	22nd.	„	20th.	„	2nd.	„	31st.
„	23rd.	„	21st.	„	3rd.	Jan.	1st.
„	24th.	„	22nd.	„	4th.	„	2nd.
„	25th.	„	23rd.	„	5th.	„	3rd.
„	26th.	„	24th.	„	6th.	„	4th.
„	27th.	„	25th.	„	7th.	„	5th.
„	28th.	„	26th.	„	8th.	„	6th.
„	29th.	„	27th.	„	9th.	„	7th.
„	30th.	„	28th.	„	10th.	„	8th.
July	1st.	„	29th.	„	11th.	„	9th.
„	2nd.	„	30th.	„	12th.	„	10th.
„	3rd.	Dec.	1st.	„	13th.	„	11th.
„	4th.	„	2nd.	„	14th.	„	12th.

Date of Service.	*Kidding Date.*	*Date of Service.*	*Kidding Date.*
Aug. 15th.	Jan. 13th.	Sept. 25th.	Feb. 23rd.
„ 16th.	„ 14th.	„ 26th.	„ 24th.
„ 17th.	„ 15th.	„ 27th.	„ 25th.
„ 18th.	„ 16th.	„ 28th.	„ 26th.
„ 19th.	„ 17th.	„ 29th.	„ 27th.
„ 20th.	„ 18th.	„ 30th.	„ 28th.
„ 21st.	„ 19th.	Oct. 1st.	March 1st.
„ 22nd.	„ 20th.	„ 2nd.	„ 2nd.
„ 23rd.	„ 21st.	„ 3rd.	„ 3rd.
„ 24th.	„ 22nd.	„ 4th.	„ 4th.
„ 25th.	„ 23rd.	„ 5th.	„ 5th.
„ 26th.	„ 24th.	„ 6th.	„ 6th.
„ 27th.	„ 25th.	„ 7th.	„ 7th.
„ 28th.	„ 26th.	„ 8th.	„ 8th.
„ 29th.	„ 27th.	„ 9th.	„ 9th.
„ 30th.	„ 28th.	„ 10th.	„ 10th.
„ 31st.	„ 29th.	„ 11th.	„ 11th.
Sept. 1st.	„ 30th.	„ 12th.	„ 12th.
„ 2nd.	„ 31st.	„ 13th.	„ 13th.
„ 3rd.	Feb. 1st.	„ 14th.	„ 14th.
„ 4th.	„ 2nd.	„ 15th.	„ 15th.
„ 5th.	„ 3rd.	„ 16th.	„ 16th.
„ 6th.	„ 4th.	„ 17th.	„ 17th.
„ 7th.	„ 5th.	„ 18th.	„ 18th.
„ 8th.	„ 6th.	„ 19th.	„ 19th.
„ 9th.	„ 7th.	„ 20th.	„ 20th.
„ 10th.	„ 8th.	„ 21st.	„ 21st.
„ 11th.	„ 9th.	„ 22nd.	„ 22nd.
„ 12th.	„ 10th.	„ 23rd.	„ 23rd.
„ 13th.	„ 11th.	„ 24th.	„ 24th.
„ 14th.	„ 12th.	„ 25th.	„ 25th.
„ 15th.	„ 13th.	„ 26th.	„ 26th.
„ 16th.	„ 14th.	„ 27th.	„ 27th.
„ 17th.	„ 15th.	„ 28th.	„ 28th.
„ 18th.	„ 16th.	„ 29th.	„ 29th.
„ 19th.	„ 17th.	„ 30th.	„ 30th.
„ 20th.	„ 18th.	„ 31st.	„ 31st.
„ 21st.	„ 19th.	Nov. 1st.	April 1st.
„ 22nd.	„ 20th.	„ 2nd.	„ 2nd.
„ 23rd.	„ 21st.	„ 3rd.	„ 3rd.
„ 24th.	„ 22nd.	„ 4th.	„ 4th.

Date of Service	*Kidding Date.*	*Date of Service*	*Kidding Date.*
Nov. 5th.	April 5th.	Dec. 4th.	May 4th.
,, 6th.	,, 6th.	,, 5th.	,, 5th.
,, 7th.	,, 7th.	,, 6th.	,, 6th.
,, 8th.	,, 8th.	,, 7th.	,, 7th.
,, 9th.	,, 9th.	,, 8th.	,, 8th.
,, 10th.	,, 10th.	,, 9th.	,, 9th.
,, 11th.	,, 11th.	,, 10th.	,, 10th.
,, 12th.	,, 12th.	,, 11th.	,, 11th.
,, 13th.	,, 13th.	,, 12th.	,, 12th.
,, 14th.	,, 14th.	,, 13th.	,, 13th.
,, 15th.	,, 15th.	,, 14th.	,, 14th.
,, 16th.	,, 16th.	,, 15th.	,, 15th.
,, 17th.	,, 17th.	,, 16th.	,, 16th.
,, 18th.	,, 18th.	,, 17th.	,, 17th.
,, 19th.	,, 19th.	,, 18th.	,, 18th.
,, 20th.	,, 20th.	,, 19th.	,, 19th.
,, 21st.	,, 21st.	,, 20th.	,, 20th.
,, 22nd.	,, 22nd.	,, 21st.	,, 21st.
,, 23rd.	,, 23rd.	,, 22nd.	,, 22nd.
,, 24th.	,, 24th.	,, 23rd.	,, 23rd.
,, 25th.	,, 25th.	,, 24th.	,, 24th.
,, 26th.	,, 26th.	,, 25th.	,, 25th.
,, 27th.	,, 27th.	,, 26th.	,, 26th.
,, 28th.	,, 28th.	,, 27th.	,, 27th.
,, 29th.	,, 29th.	,, 28th.	,, 28th.
,, 30th.	,, 30th.	,, 29th.	,, 29th.
Dec. 1st.	May 1st.	,, 30th.	,, 30th.
,, 2nd.	,, 2nd.	,, 31st.	,, 31st.
,, 3rd.	,, 3rd.		

CHAPTER NINE

KID REARING

You should make up your mind what you are going to do with the kids before they are born, so that you can begin as you mean to continue. Otherwise you may have a lot of trouble with a fretting nanny and consequent loss of milk. Some goats are devoted mothers and very hard to wean from their young ones, while others are thankful to see the back of them in about a month. Some nannies do not even take an interest in their offspring when they are born and will attempt to injure them, butting them with their heads and pawing them, and the sharp hooves can cause severe injury. Again, a goat that has always been milked by hand and has never suckled kids will probably strongly object to having them near her and will kick or trample them and butt them off their feet, if they attempt to suck. Some nannies can be very nasty to their kids. However, the average goat is devoted to them, and it causes a lot of trouble if you leave them with her for a week or two and then suddenly take them away.

If you have decided, in advance, that you will let her rear any nanny kids she has, well and good, if she is a good mother. This system has its advantages and, also, its disadvantages. You are saved the trouble of hand-feeding, and the kids get their food, uncontaminated and at the right temperature, whenever they require it. If you are away from home a good deal and can only attend to your goats at certain hours, or if you have to leave them to unreliable or rather youthful helpers, then it is certainly best to leave the kids with their mother. But, in this case, you should let her keep them for at least a month, and preferably six weeks, and the weaning should be done gradually.

In the meantime, you may not be getting all the

milk you would like. The kids will take what they want, and you will have what is left. Some people attempt to right this by taking what they want first and leaving the rest to the kids ; but this will not do, because few goats will let their kids suck them dry, and, if the nanny is not properly stripped twice a day, her yield will diminish. Consequently, if your nanny is rearing kids, you have to let them take all they want, and you strip her every morning and evening and make do with what you get. This is the main disadvantage of the natural form of kid-rearing. However, it does not last long, and, as a mother-reared kid generally learns to eat hay and meal sooner than a hand-reared one, you may find yourself getting more, instead of less, milk towards the end of the suckling period.

If you find the kids troublesome, when you go to do the stripping, you will have to tie them up or put them in a little pen made by tying hurdles together, near the mother's stall. Do not take them right away, or even put them outside the door, or the mother will probably get in a stew and refuse to let you touch her. Even if you wedge her in a corner and make her stand, you may not get the milk, for they can hold it from you so that you think there is none. This will be dealt with more fully in a chapter on milking. Stripping, of course, means getting the last of the milk from the udder; and it is in these last drops that most of the cream is found.

Remember, however, that a goat must never be stripped dry during the first four days after kidding, or she may collapse from calcium deficiency.

Kids that are running with their mother can go out with her when only a few days old, if the weather is mild. If, however, it is too cold or wet for them, the mother should go out for a short period of grazing each day, from the time they are about three or four days of age. At first, she will not like leaving them, and she should not be left out for more than an hour or so. But, as she finds they are always still there on her return, she will get used to leaving them and they will get used to her going out, and the pathetic calling back and forth will cease. She can then stay out for longer periods, until, if you can spare the time, you can leave her out all day, giving the kids a bottle of milk during her absence.

Whether you do this or not, you should begin weaning the kids when they are a month old, by taking them away from her at night, after she has fed them, and giving them back to her in the morning. By this time, they should be eating a certain amount of solid food; and, as their appetites for this increase, they will be able to do without the midday milk and just go to their mother twice a day only, in the morning and evening. In due course, this is reduced to once a day, and eventually they are weaned altogether.

This is the natural method of kid-rearing. It saves you a lot of trouble, but it may leave you short of milk. Also, it quite definitely leaves you without a record of the goat's yield. If you want all the milk you can get, or if you are recording, or if you find yourself with a nanny that objects to being suckled or treats her kids badly, then you should hand-rear them. But this is not a method to attempt unless you have the time to do it properly, also the patience and perseverence to stick to it conscientiously. Hand-rearing is a great tie, and you must be very careful to feed regularly, in correct quantities, and at the right temperature, and to observe strict cleanliness in the handling of the milk, bottles, and teats. In fact, it is just as bad as having a human baby to tend! The only difference is that it only lasts a short time. However, weigh the matter carefully before you decide on hand-rearing, for, though you can rear fine kids by this method, you can, also, ruin them by carelessness. It all depends on you!

If you have definitely decided, in advance, to hand-rear the nanny kids, you now have to consider whether you will take them away at birth or whether you will leave them with the mother for a few days. Some goats do not mind you taking them and seem quite thankful to be alone with the familiar human again. But most of them love their babies dearly for a few weeks, even if it does sometimes wear off rather quickly! You can leave mother and kids together for three days and then take them away while she is out grazing, and she will not make much fuss about it. It is when she has had them for two or three weeks that separation upsets her, aggravated, of course, by the fact that lusty kids of this age make a loud outcry for their mother. If they are not taken away after three days, they should be left for the full period of a month to six weeks.

The normal, healthy kid is on its feet very soon after it is born and looking for a teat within an hour or two. The stronger ones generally help themselves without much trouble, but later-born kids, especially if they are small, may get pushed away or find it difficult to get the teats into their mouths. Watch them, when they first begin, and help any that seem to need it. Hold the kid's head upwards with one hand, and put the teat to the mouth with the other. Do not squirt milk into the mouth. Let the kid grasp the teat itself. It will soon learn that it can retain its grip by suction and that this causes the delicious, warm milk to flow. It is seldom that a second lesson is needed, though it may be necessary to haul off a big, greedy one for a spell, while a little sister gets a chance.

The fluid that comes during the first three days is not proper milk but colostrum. It is yellow in colour and curdles if brought to the boil. It is not fit for ordinary household purposes, though some country people look upon it as a delicacy and are very pleased if they can get hold of some " beastings ", as it is called. However, to most people it does not look at all appetising, for it contains mucus, but it is very necessary that the kids should have it, for it is Nature's aperient for the new-born. If, through the loss of the mother, it is not available, they should have a small teaspoonful of castor oil before their first feed. If you are not keeping the kids and have the colostrum to spare, or have more than they need, it will be enjoyed by the dogs. By the fourth day, the milk should be normal and fit for all usual purposes. Frothing when it is milked is the sign to go by, but, if in doubt, boil a little. If it boils properly and does not curdle, it is all right.

It is at this time that the kids that are to be hand-reared are taken from their mother and put in the shed where they are to be reared. Take them when she is out for her morning graze, and put them, if possible, right out of sight and hearing of the mother. The shed should be light and well-ventilated, but it must not be draughty and should be reasonably warm. A box should be provided for the kids to sleep in, and this should be well lined with soft hay. Straw does for the floor of the shed. Be careful not to leave tools or movable fittings in the place, for the kids frisk about when only a few days old and accidents often occur through playfulness. The barer the shed the better, and there is no need for anything in it but the sleeping-box and hayrack, also the manger, if it is a fixture.

You now have to hand-feed your kids every few hours, and you must be very regular and punctual about it, or they may have digestive trouble. And a kid can very easily die of this. Four-hourly intervals are best, as with human babies, and five feeds a day are needed for the first fortnight. After this, they have four feeds a day, until they are six weeks of age. By then, the worst of it is over, for they will be eating a certain amount of solid food; and, from six to twelve weeks, only three bottles a day are given. For the next two months, they have two bottles a day, and then, for another month, one bottle a day. By this time, they are six months of age and should need no more bottles, though some people continue to give one bottle a day through the worst of the winter, if they have six-month-old kids on their hands at this time. Much, of course, depends on the weather and the time of year at which the kids were born. Some people do not carry on with bottles for as long as six months, especially if the kids are getting plenty of young, tender greenstuff; but it is generally advisable to do so, especially if you are hoping to rear heavy-milking goats. You can, however, get your kids to drink from a bucket during the latter part of the rearing period, and this saves messing about with bottles and teats.

Some people hand-rear their kids with a bucket from the very beginning, and it is, of course, the customary thing with calves. The required amount of milk is put into a bucket, (at rather more than blood heat, to allow for the quick cooling that takes place when a bucket is used), and the hand is put in the milk, palm upwards, with one milky finger showing. The other hand is used to show the kid where to look for nourishment,—and quite a lot of strength is needed, at first, to convince it that it is making a mistake in turning its little head upwards! Once it learns to drink the milk by sucking the finger, it is not long before it drinks from the bucket without a finger, and feeding becomes a rapid business. But that is actually the drawback to the system, for they drink too fast for good digestion. They get their living too easily and take down their milk without the amount of saliva that would be induced to flow by the process of sucking. Also, they swallow air with their milk, as they gulp it down greedily; and this, together with the lack of saliva, causes indigestion and tends to make the kids "pot-bellied". It improves matters if you make them stop drinking halfway through and take a short breather, but it is by no means the ideal way of rearing kids. However, when they get to the stage of only having two such meals a day, there is no reason why they should not learn to take it from a bucket, as then they have solid foods to digest and the milk is only a part of their diet. In fact, it is a very good thing to teach them to drink from buckets while there is still their beloved milk to tempt them, as then they are more likely to take to water.

For baby kids that are to be reared by hand, it is certainly best to begin with bottles and teats. Ordinary feeding bottles, as used for human babies, are right for the first week ; and pure, undiluted goats' milk should be given at blood heat,—which, in the case of the goat, is 103 degrees, Fahrenheit. The warming can either be done by standing the bottle of milk in a jug of hot water, (as most mothers do for their own babies), or by warming it in a saucepan. No hard-and-fast rule can be laid down as to how much milk each kid should have, as their appetites vary, but they should have what they will take willingly, without being pressed.

Very soon the kids will want more than one baby's feeding-bottle full at a time, and, in due course, you can put them on to ordinary wine bottles, (which hold about a pint and a quarter), fitted with lamb teats. They will not need them full at first. Go by each individual kid's appetite, and never force it to take more than it wants. Also, never offer what one kid has left to another that has already been fed. Too much milk may cause scours, and no kid should ever have more than a wine-bottle full at one meal. And, apart from the risk of scours, it is wasteful giving kids more milk than they need and tends to put them off eating solid foods, which they certainly ought to have when they are a month old. There is definitely such a thing as giving too much milk, so that the kids satisfy themselves by drinking and do not bother to eat.

When rearing kids by hand, one must observe strict cleanliness in the matter of handling the milk and sterilising bottles and teats. Also, the teats must be renewed pretty often, as, when they get soft, the kids get their living too easily. Ask any mother about this ! It certainly means a good deal of trouble, and you should not attempt to hand-rear kids unless you realise what it involves and intend to make a good job of it. But the system has its advantages, and not the least is that, when the kids are a month old, you can begin to use milk substitute or calf meal, instead of giving them all milk. This comes very much cheaper, for the cost of a good milk substitute works out at about threepence a gallon.

Substitutes can be very good, but they must be used with discretion. Do not keep chopping and changing. Decide on one brand and give it a fair trial. If you find it unsatisfactory, try another and give that a fair trial. If you keep trying this and that, without giving the kids a chance to get used to anything, you will upset their digestions.

There are various proprietary foods advertised for use as milk substitutes, some of them made especially for kids; and your own food merchant probably sells No. 1 Calf Meal, which is what farmers use for rearing calves. Decide what you will use, and make it strictly according to directions. Calf meals often contain linseed meal, and this must be well boiled before use, as it is liable to generate prussic acid when wet. If it is fed in the dry form, it is quite safe. But, if it is made wet, it must be thoroughly boiled, (and without a lid), so that the poisonous properties can be dispersed.

Do not feed calf meal or milk substitutes until the kids are a month old. Up to this time, they should have pure, undiluted goats' milk. Cases have been known of kids being reared quite well on substitutes from a much earlier age than this; and, if you should have the bad luck to lose the mother, or if you buy an orphan kid, it is the only thing to do. But they do not seem to make quite such good animals as they would if given pure milk for the first month. After this, the substitute should be introduced gradually, giving part meal and part milk and gradually increasing the amount of meal and decreasing the amount of milk, until eventually they are having all meal and no milk.

In the meantime, the kids should be learning to eat solid food. Just when they will begin depends a great deal on the individual. Some begin to nibble bits of this and that when only a fortnight old, and the average kid eats quite well when a month old. This gives some people the erroneous impression that they no longer need milk, and you have to beware of people who seek to buy kids from you with the intention of putting them out to graze and leaving it at that. I always ask prospective customers if they realise that a kid under six months of age needs milk or a milk substitute, and it is amazing to note the number of people who think a kid can be tethered on a lawn and left to fend for itself.

Their digestions cannot cope with solid food alone, while they are still babies, but they certainly ought to learn to eat hay, and tender greenstuff, and bran. They will, also, grub up earth and eat it, and this should not be looked upon as a depraved taste. Other grazing stock do the same, both in the wild and domesticated state. They are searching for minerals that their instinct tells them they need, and you should see that all goats, at all ages, have access to rock salt and mineral licks.

As the kids learn to eat more freely, try them with a little finely broken cake. They have sharp little teeth and like to use them, and, once they have developed the habit of eating cake, you can give them their calf meal in the form of dairy nuts. But still let them have the various cakes and meals that they will have to eat as adults. Kids only eat very small quantities at a time, but they should be taught to eat a variety of foods, so that they do not grow up finicky and difficult. They should, also, be taught, from quite an early age, to drink water, and this they are more likely to do if they have access to salt.

Kids that have to do their growing through the winter months will grow more slowly than their older sisters. There is not much nourishment in wild greenstuff at this season of the year, and they will have to depend on what you can grow for them, together with roots and hay. At first, the roots will have to be pulped or cut up for them; but, once they have the teeth for gnawing their roots, I believe in letting them do so, as it helps to develop the jaw and teeth and, also, obviously gives them pleasure.

If you only have a few goats, you will doubtless be able to provide them with enough greenstuff from your garden or allotment, but, if you have to grow special crops for them, you might consider lucerne, or alfalfa, as it is sometimes called. Once established, this hardy crop can be cut three times a year, in almost any weather, and it should stand for four or five years, or more. It is greatly enjoyed by goats of all ages and is better suited to kids than kales and cabbage, and it contains a good percentage of protein, both in the green and the dry state. It is richer in protein than good-quality lawn-mowings, which are considered a valuable protein food. It is a particularly useful crop for places where the rainfall is low.

When introducing a new food to goats of any age, and most particularly to kids, be very careful to do so gradually. A sudden change, or a whole complete meal of something they have not had before, (or have not had for some time), is liable to cause scouring. If this is severe, it causes great pain, and little kids cannot endure great pain and are liable to die of it, although there is no actual illness. It is a great pity to lose healthy kids through a sudden bout of diarrhœa, especially when it is just due to thoughtlessness.

Little kids are exceedingly playful and very amusing to watch, and it is a good idea to provide them with some simple " toys ", as plenty of exercise helps them to grow into lusty, well-developed animals. A favourite " toy " is an old tree-stump, or a strong box. See that it is absolutely firm and level, so that it cannot easily overturn and trap a kid beneath. It will not be long before this is discovered, and your kids will soon be playing that age-old game beloved of all human kiddies,—" I'm the King of the Castle! " Two or three will leap up and start heading each other off, and they land daintily on their small feet and never seem to come to any harm.

A similar game will soon start if you nail a plank to a couple of boxes in their pen. A kid will jump on at either end, and they will meet in the middle and put their heads together and push, and push, and push,—until one goes overboard! You can waste no end of time, watching your kids skylarking!

DISBUDDING

Before we leave the subject of kid-rearing, there is the question of disbudding to consider. This should always be done, unless you have to put your goats out where they may need the protection of horns. But a horned goat is a nuisance in her stall, for she spends her spare time doing fancy-work with them! Any goat, without horns, can be hard on her hayrack, buckets, etc.; but, with horns, they can pretty well bust up the place! Also, they must not be put with hornless goats, or they have an unfair advantage; and they often give their owner a nasty whack,—with the best of intentions,—simply by trying to chase off a fly or scratch a tickle on the back, when he, or she, is milking!

Disbudding is quite easy, but it is a thing that needs practice. Your first attempts may not be entirely successful, and the horns may still grow a little. However, it is definitely one of those cases where practice makes perfect. It should be done when the kid is from two to five days old, otherwise it is not likely to be effective. A stick of caustic potash, or a proprietary disbudding stick made for kids and calves, is used, and you ought to have an assistant to take the kid on his, or her, knees and hold the head firmly and steadily between the hands. It is difficult to manage properly alone, for, though you can nip body and legs between your own knees, it is very hard to keep the kid's head still while you use the disbudding stick. The method is as follows:—

First clip the hair from around the two horn-buds,

A pair of curved manicure scissors is excellent for the purpose. You can easily find the horn-buds, even though they do not feel hard to the touch for the first few days, for each one is encircled by a curl of hair. (In a naturally hornless kid, there are no curls, and the forehead is slightly rounded, instead of being flat.) In the centre of these curls, you will find a very small patch of clear skin. This is the horn-bud, and it becomes an eminence in a matter of hours after birth. Clip the hair away carefully, and smear vaseline round the horn-buds. This is to prevent the caustic substance from running on to the tender flesh and burning it.

Wrap a piece of brown paper or tinfoil round the end of the potash or disbudding stick, (so that you do not burn yourself), and moisten the end very slightly, or else slightly moisten the horn-buds. Not much moisture is needed, as exposure to the air makes the disbudder still wetter. Some people use saliva to dab the horn-bud, and some use a piece of damp blotting-paper on the disbudding stick. Now rub the horn-bud for a quarter of a minute with the disbudding stick, taking care not to rub too hard or to go beyond the circle of vaseline. An area about the size of a half-penny should be treated. Let the kid have a few minutes' rest and then repeat the process. Three or four applications may be necessary, with a few minutes' rest between each. Be careful not to rub so hard that the skin is broken and blood appears, and remember that it is the very centre of the horn-bud that needs the most attention.

By now, the circle should look damp and rather blackish, and the centre of the horn-bud should be soft and show signs of blistering. Dab the excess moisture from the circles, and put the potash or disbudding stick back in its tube and stopper it tightly, for it dissolves quickly when exposed to the air. The operation is now over and the kid can be allowed to run about, but it should be watched carefully for a little while, to see it does not rub its head against things. The treatment does not cause pain, if properly done, but it seems to worry the kid for a short time,—until it forgets all about it.

It is very necessary to see a disbudded kid does not get its head wet for some days afterwards. Do not let it go out in the rain or get milk on its head when feeding. This is a difficult problem with mother-reared kids, as the mother sees something amiss about the head and wants to lick it. If you tie anything over the head, she will certainly try to get it off and will almost certainly succeed. Consequently, it is better to take the kids from their mother when three days old, as advised earlier in this chapter, and rear them on the bottle after disbudding.

It is not much use trying to disbud a kid after it is a week old. It *has* been done,—in fact, cases have occurred when disbudding has been successful as late as ten days,—but there is always considerable doubt about it, when it is left until the horns have become more than mere eminences. Of course, there is no harm in trying. A partial, or ineffective, disbudding does not harm the kid. It simply means you do not get the results you want. Bad disbudding sometimes results in deformed horns growing.

Once you let the horns grow, there is little you can do about it. Within a few weeks, you will find the kid is actually scratching the mother's udder and making small wounds that bleed. Kids punch their mothers unmercifully with their hard heads, to make more milk come down, and, if they are growing horns, they should be weaned promptly, before they do harm.

Horned kids can be put together, but no hornless kid must be put among them, for their pet game is head-shoving and the little horns can be very painful. Very soon, they will be using these troublesome little tools for all sorts of purposes. They not only shove and butt with them but, also, use them for tipping things over and banging about in general. Later on, as they become long and curved, the goats use them for scratching their backs, flicking off flies, and scraping the walls of their sheds. The latter game has no apparent point. It is merely a case of wiling away the time doing fancy work, just as children pick away the wall-paper as they lie in bed! In short, horns on a domestic goat are very undesirable, and you generally find that hornlessness, (whether natural or due to dis-budding), adds anything from ten to twenty shillings to the value of a goat.

Sawing horns off an adult is illegal in this country, though apparently practised in America. There is, also, another method practised, which appears to be painless, merely involving the cutting off of the blood supply to the horns. Whether or not this would be allowed in Great Britain I cannot say, for I have never heard of it being done, and readers wanting to try it should first satisfy themselves on this point. At the time of writing, it has only recently been reported, and it is not yet known if the horns grow again, after having been removed. The method is simply to put a narrow, elastic band round the base of the horn, nip-ping up a tiny ring of skin at the same time. Narrow bands must be used, as it is found that wider bands ride up the horn and so do no good. The narrow band grips tightly, and the blood supply is immediately cut off from the horn, so that, within a few months, the horn falls off, of its own accord. This sounds harmless enough,—much like putting a ligature on a wart,—and certainly far less severe than many painful processes which are strictly legal, such as docking, dubbing, and castrating. However, readers should bear in mind that this description of the method is only given in the form of a report of what is done in another country, and it yet remains to be seen how it will be accepted over here.

Horns, which are meant to be a form of protection to their owner, can be a danger to the possessor, as well as a nuisance to the goat-owner. Doubtless the reader remembers the Old Testament story, of the ram which Abraham found "caught in a thicket by his horns". Many horned animals must have been caught in the same way, before and since, and the author has heard of a quite recent case, in which a domestic goat caught one of its horns in the bars of an iron rack and wrenched the horn off, in its attempts to get free. This is a very nasty accident, for the soft core is left exposed and bleeds a good deal, and the animal must suffer much pain.

When horns are injured in this way, it may be necessary to have them removed, but this must be done by a veterinary surgeon and under a total anæsthetic, for it is a major operation and a very ticklish one. Even when properly done by a qualified man, it is by no means certain it will be successful. There may be bleeding from the nose, due to the injury to the frontal sinuses. There may even be septic inflammation of the sinuses, with the formation of a great deal of pus, and perhaps death from meningitis. Needless to say, it is illegal for any unqualified person to attempt such an operation.

There is one other matter that may need attention about the time that kids are disbudded, and that is the navel cord. Ordinarily, this is ruptured at a suitable length at the moment of birth, and the rest dries up and falls off within a few days, leaving the navel clean and healthy. Sometimes, however, the cord is left too long, and there is the risk of it being trodden on by the kid itself, or by other kids, or by its mother. In such a case, it should be trimmed off with a sharp pair of scissors, which have been sterilised in boiling water. Do not do this on the day of birth, unless an exceptionally long cord is left. Wait a day or so, for it to dry naturally, as this prevents the entry of germs. Then trim it off to within about an inch and a half of the body. Do not cut into the part where blood-vessels are still active.

CHAPTER TEN

MALE GOATS

So far, in writing of the rearing of kids, I have only mentioned nanny kids. As a rule, the best thing to do with billy kids is to use them for table. The demand for billy goats is very small, and you only need to work out the average cost of rearing a kid to find out that it simply is not worth while bringing up a billy, unless you need him for yourself or can be pretty certain of getting a really good price for him. And, even then, you run the risk of disappointment, for a surprisingly large number of billies turn out to be no use for stud work. Why this should be so common a failing among goats it is difficult to say. Some people think it is due to breeding for hornlessness, so producing an effeminate type of male and increasing the proportion of hermaphrodites. However, whatever the cause, the fact remains that many apparently sound and desirable billies are not stock-getters ; and, once you have reared a billy to breeding age, there is nothing for it but to sell him to a knacker, (for dog meat), if he will not breed.

There is not a great demand for billy goats, unless of quite exceptional quality and proved stock-getters. There are so many excellent males at stud under the Stud Goat Scheme, in addition to the many more outside the Scheme, and a vigorous male can serve as many as seventy nannies in one breeding season. Consequently, it is not difficult, in normal times, to send one's nannies away to a really good billy at no great distance ; and many goat-keepers prefer to do this, rather than keep a stud male of their own. There is no denying that they have a quite distinctive and very penetrating odour, and their displays of affection to you are liable to prove quite embarrassing. A male goat should have special quarters, right away from the nannies, and you should wear a special overall when you attend to him. Otherwise, fastidious members of the family may nip their noses when you enter the house!

However, if you are used to the ways of livestock and have plenty of room, you may like to keep your own stud billy. It saves you the trouble of sending the nannies away, and a good male should earn a tidy sum by stud work. He should be a really good one. Do not keep a certain billy kid just because he looks a beauty. If you plan to rear a stud billy, get the best blood you can afford on both sides. You may have to wait several years and make several such good matings, before you get what you really want; but I think this is better, in the end, than buying a stud male. A bought animal can be seen at once for what he is, and you can insist on having a guarantee of money refunded, if he fails to prove a stock-getter, and you can use him at once. But, against all this, there is the undeniable fact that a stud male goat, like many other breeding males, can be "an ugly customer". They are incredibly strong and weigh about a couple of hundred pounds, and, if one just decides he does not like you, you have a troublesome problem on your hands.

But a stud male that you have reared from a kid will almost certainly grow up very attached to you. They are loving creatures and will display their affection often to an embarrassing extent, and they expect a similar display from you. So what you might not be able to accomplish by force, you can easily achieve by playing on his affection. By this, I do not mean that you should play larks with him. Do not play with male kids being reared for stud work, however much you may be tempted to do so when they are small and amusing. The secret of managing a stud billy goat without trouble is a combination of firmness and affection. Never let him feel his own strength. Never tempt him to try it. Let him wear a headstall always, and teach him, from babyhood, to walk sedately and obey you. But pet him, by all means. Rub his neck and scratch him behind the ears, and he will obey you from sheer affection. Many a slight and delicate woman manages her own stud billy goat and finds him as gentle as the proverbial lamb.

A billy kid can be reared with the little nannies, at first, and treated just as they are. Let him run and play with them; for the more exercise he gets, the better animal he will make. But do not let him remain with them after he is three months of age, if it is the breeding season, for there is the risk that he might get one of the little nannies in kid, even as young as this. Do not give him too much milk. Since they are used for stud work when six months old, they are usually weaned off milk rather earlier than the nannies; and the more hay, etc., you can induce a young billy to eat during his growing period, the better animal he should be. Let him have plenty of concentrates,—as much as he will eat eagerly,—and encourage him to drink water, and see he always has access to rock salt and a mineral lick.

When you have to take him away from the nannies, he will have rather a lonely life, and lonely males often develop disgusting habits. Let him have spells out-of-doors whenever you can. A male goat can be tethered, just like a nanny, but it is hardly safe to do so during the breeding season, for they can get away from almost anything, if they choose to exert their strength. The best thing to do is to let the male have a strong yard attached to his shed, and let him feed and exercise in this. Give him some branches to bark and some strong and harmless "toy" on which he can expend his spirits, and let him have a change in the fields when the nannies are in their stalls. And do not forget to give him a name and call him by it. A special name, kept for him alone, is such a bond of affection between you that you soon have the poor, lonely fellow quite literally at your beck and call!

When he is six months old, he should be well grown

and well developed, and you can then use him lightly for stud work. Do not let him overdo it, for he is still a kid and is still growing. Individuals vary a good deal, but the kid's own behaviour towards nannies is a safe guide. If he is at all slow in his approach, you are probably overdoing it. Do not, however, mistake shyness for slowness. Many nannies begin by butting their suitors, and a young billy may try "appeasement", paying gentle attentions until the nanny becomes interested. But, if he shows lack of eagerness when he really has the chance, he should be given a rest from stud work.

During the breeding season, a stud male goat will go off his food and lose condition, often quite badly. This is usual, and there is little one can do about it, beyond tempting him with titbits of anything one knows he particularly likes. All goats love bread and cake,—the latter of the variety found on our own tea-tables!—and any oddments of this sort will be welcomed by the lovelorn swain. He will, also, enjoy the papers that are found round cakes! Ivy is another thing that is seldom refused. Do anything you can to keep a male goat eating, especially when he is a kid; but do not worry unduly about the loss of condition, for they pick up again when the breeding season has ended.

If you are keeping a stud male goat, you must know how to manage him. If you have a suitable yard, you can put nanny and billy in this for mating. But, if you have to put them in a field, you must retain some control over them, as mating is often preceded by what appears to be a fight. Doubtless it is only the equivalent of the slapping on the back in which some humans indulge, but it is pretty vigorous and may result in injury. Some nannies will stand for service at once, but many like to have a run for their money, so to speak, and lead the male a dance first. So you should have two long cords, with spring hooks on the ends. Fix one on each animal and lead them out, holding them close to the heads, until you have reached the chosen spot. Then let them out, keeping a firm hold on the ends of the cords, (which are best with a loop for the hand), and see how the goats behave with each other. Do not let the male frighten a nanny, by sudden attentions before she is used to him. Maidens are often nervous. And do not let a mature and experienced nanny bully a young billy. With a cord attached to each, you can draw them apart at will ; and, if there is any roughness, it is best to let them walk about a bit and get used to each other, before attempting to mate.

It is always helpful to have someone else at hand, when there is a nanny to be served. If the goats do not suit each other for height, one or other may have to be raised, by pushing a truss of straw, (or something similar), under its hind legs. This is rather difficult to manage alone, as the animals may move and change position just when you have put the truss in place. But, with a helper, you can guide them into a suitable corner, or against a wall, and stand at the head of the nanny, while your helper puts the straw into position and leads the billy up from behind. Do not, however, do this until you are sure they have finished their preliminary canter and are ready for mating.

If you accept nannies from other people, for service by your billy, you should have proper accommodation for them, as some will not mate at once. Also, those that come by rail will, at the very least, have to be kept overnight. Feed and milk them, as though they were your own, but keep them in separate quarters from your own goats. One never knows when an animal may carry infection or have picked up a chill on the journey. Look over every visiting nanny, before you let her come into contact with your billy. See she is clean and healthy and that she has no unusual discharge. See, also, that she is properly in season. You are within your rights in refusing to give service to any nanny that you consider in an unfit or undesirable condition, and you are not bound to assign a reason.

Remember, however, to return the stud fee, if you do not wish your billy to give service to any particular nanny.

Although there is no legal claim, it is usual to give a second service free, if the first should prove ineffective. This point should be made quite clear, before you accept a nanny. It is, also, quite common to give an absolutely free first service, if you are trying out a kid and do not know if he is a stock-getter. In such a case, you should tell your client that the kid is untried and that you will give a free service, if the client will let you know promptly if it has been effective. If you have no nannies of your own due in season just then, this may save you valuable time and be well worth the loss of the stud fee.

When the breeding season is over, there is no reason why the male should not go out with the nannies, so long as he is gentle and well-behaved with them. Some billies will guard the herd devotedly and take a fatherly interest in the washing of the kids; and this is an advantage, if your goats run free. Trespassing boys and dogs will think twice before interfering with a herd guarded by a billy. Some people allow their stud billies to remain with the nannies all the year round and mate at will; but the disadvantage of this is that you are never sure whether or not a goat has been served, unless her figure shows it, and you know next to nothing of the mating dates.

There is still another purpose for which a male goat may be kept, and that is as a draught animal. They are castrated when about three months of age, and they grow into large, handsome, tractable animals of great strength. They are a delight to children and can be harnessed into little traps, just like small ponies. Also, they can be really useful on such a place as a poultry farm, where a fairly small, but strong, animal will more than earn its keep, in drawing a little cart from pen to pen, and taking around the week's supply of mash, and so on. Neutered goats do not have any odour.

However, for the vast majority of billy kids, even though they may be very well bred, there is nothing for it but to kill them and have them for table, or use them for dog meat. Do not let them go as children's pets, under the impression that it is kinder than killing them, unless you are quite sure they will be well and properly reared. Many people take kids as pets for their children, imagining that they can find their own keep on the lawn, or on a bit of rough land somewhere. The very idea of giving them milk amazes them. And, even if they do undertake to give them milk and rear them properly, it is ten to one they will be forgotten and neglected when the children tire of them. It is far kinder to give unwanted kids a quick and merciful end.

Any time from a month up to six months of age, a male kid can be used for table. Actually, I have had them when only a fortnight old, but they are rather tasteless and inclined to be " rubbery " when so young. If it is decided to kill the billies at birth, it is best to use them for dog food. From two to four weeks, they can be made into stew, for human consumption, if plenty of herbs are used, to give flavour. But, from a month upwards, they can be roasted, like lamb, and they are delicious. There are four joints,—two legs and two shoulders,—and there are the head, heart, neck, and rib portions for stew, and the liver, kidneys, and brains for a delicious quick fry. Kid liver is tenderer than any liver obtained from the butcher, and there is generally a fair amount of fat in the carcass. The parts for frying can be used as soon as wanted, but the rest of the carcass is best hung to dry for a few days before eating. How long it should be hung depends on the weather and the place used. A windy place should be chosen, in preference to a still atmosphere that may be damp. I hang mine from a tree until cool.

The skin should be taken off carefully and tacked out and dried, just as rabbit skins are. They can then be dressed and make useful mats or rugs, according to size. Even the feet and lower parts of the legs need not be wasted, if you have dogs, for they are really only gristle. My dogs eat the lot, also the lungs and "innards"!

Billies should not be killed for table too near the mating season, if they are of an age to breed, as then the flesh is liable to be tainted with the male odour. If you have run on a male for stud and find it will not breed, there is nothing for it but to kill it for dog meat or sell it to a knacker for the same purpose. Knackers are to be found in most country districts, and their advertisements for old cows and horses will be found in the county papers. A neutered goat can be run on for as long as eighteen months and can then be cured, like pig meat, and made into bacon and ham.

As regards the actual killing, very young kids can be killed quite easily, like rabbits, with a sharp blow behind the ears. For older kids, it is necessary to use the knife. A butcher will generally kill, and skin, and divide up a kid for a customer, and, even during the time of restrictions, due to rationing,—still in force, at the time of writing,—this was in order. There are not, (and were not, even during the war), any regulations or restrictions regarding goat products, (except that the butter and meat are price-controlled), and you can kill a kid, or have your butcher kill it, and do what you like with the meat. A butcher kills a kid as he kills a lamb, piercing the jugular veins by putting a sharp knife through the throat, behind the jaw, going in at one side and out at the other. It is quickly done, and the kid has bled to death in a few moments. Personally, I prefer to stun the kid before using the knife, but butchers prefer not to do so, saying they bleed better when not stunned.

If you have an adult unwanted goat, the best thing is to shoot it, if you do not want to send it to the knacker, who may not be so kind. During the war years, I bought cheap, unwanted goats, often in poor condition, for my pedigree Cairn terriers, the dog meat position being often very difficult. Having shot the goat, I always took off the skin, before cutting up the meat for the dogs; and it is surprising what lovely rugs they make, even if the goats are in poor condition. Never waste a skin, even if it is off a tiny kid. They are attractive and hard-wearing, and, once dried, they can be put away to be dressed at any convenient time. So do not throw away a skin, just because you happen to be busy. Tack it out in its natural shape, and scrape off the fat as it dries, just as you do with rabbit skins. When dry, roll it up over some corrugated cardboard, put in some mothballs, and wrap it up in newspaper. Never fold a skin flat. Always roll it.

CHAPTER ELEVEN

GENERAL MANAGEMENT

All goats should be groomed regularly, about once a week. This is desirable for the good of their health, for it helps to keep the skin in good condition, stimulates the growth of the coat, and gets rid of dead hair. Needless to say, it greatly improves the appearance of the animal as well. Just as we brush and comb our own hair and that of our dogs, so we should groom our goats; and a stiff, long-bristled brush and a steel comb, such as are used for dogs, are best. Most goats enjoy grooming and will stand beautifully while it appears to be a form of stroking, but some are restive when you get to the lower parts of their legs.

Another reason for grooming is that it gives you a chance to see if the goat has collected any undesirable livestock. Like other animals, they are liable to become infested with lice or fleas,—more particularly the former,—though it is not often the case with goats that are kept in clean surroundings. You may, however, have the misfortune to purchase a goat in this dirty state, and, if you do, she must be treated as quickly as possible. Neglected goats get absolutely lousy, and I have known of cases where they have died of anæmia caused by these creatures.

The lice that infest goats are practically colourless and extremely small,—so small, in fact, that you are likely to mistake them for scurf, unless you watch very carefully. They are slow, not moving unless disturbed with a finger and then only moving slightly, and they are sometimes very numerous. All this helps the illusion that a lousy animal is merely scurfy. You need to look very carefully, in a bright light, especially down the back. You may, however, have your suspicions aroused by the fact that a certain goat is constantly licking herself. Although they sometimes use their horns to allay irritation, goats do not do much scratching. They lick the part that worries them, and, if you see those tell-tale lick-holes very often in a certain animal's coat, you may be sure it has insects.

Fortunately, since the creatures make no effort to escape, it is easy enough to get rid of them with a good insecticide. For preference, use one containing D.D.T. powder. Turn up the coat, here and there, especially along the back, and round the neck, and behind the ears, apply the powder, and rub it in well. After this, do not do any brushing or combing for at least a week. Then brush out the old powder and give a second dose. Leave this for a week, and then brush well, and the animal should be clear.

Some people like to wash the goat completely, using one of the insecticide shampoos made for dogs. If you do this, choose a fine, sunny day, with perhaps a little wind, but definitely not a cold wind. Goats generally like being washed, and you can expect them to stand steadily while you lather and rub them. Leave the suds to soak in, while you get the rinsing water ready; and, after rinsing, give the goat a final rinse with a watering can, using the rose. Rub down briskly with clean sacking, finishing with an old, rough towel, and then let the goat go out and air herself in the sunshine. Washing is a useful treatment when a goat is just scurfy, apart from lice, and it is, also, the best course when an animal comes to you with a really dirty-looking coat. White goats often look very bad, if they go out in all weathers and get no grooming. Normally, brushing and combing is sufficient to keep them looking clean and spruce, though a rub-down with a piece of clean sacking is needed, if a goat comes in wet.

If you particularly want a certain goat to look its best,—as, for instance, when you are going to exhibit it, or if you have visitors coming to view your stud billy,—you can improve the appearance by giving a light-coated goat a dry shampoo. The sort made for dogs do very well. Black and dark-coloured goats respond best to rubbing with a soft cloth after grooming. Do the rubbing the way the coat lies,—in fact, make it stroking rather than rubbing. Remember how much glossier the pet cat looks after stroking! Another dodge for making a goat look its best is to polish the hooves with brown boot polish.

The hooves must, of course, receive regular attention. How often this is needed depends on the individual and the kind of life it leads. In some cases, the horn grows much more quickly than in others, just as our own rate of nail-growth varies. Again, goats that walk much on hard roads wear their hooves down naturally and do not need them trimming nearly so often as goats that spend much time standing in their stalls, or only walk in soft pastures. The hooves must, however, be regularly inspected, as, if they are allowed to get in a bad state, foot-rot may develop.

Trimming the hooves is quite easy, if they are wet, but difficult, and sometimes impossible, when they are very dry and hard. Consequently, they should be done when the goat has been out grazing in damp grass. Alternatively, you could walk her through a really wet patch, or through a stream. I have heard of people tying up the hooves in pieces of sacking containing cow dung, but I should not recommend it, if you can soften the horn by cleaner and simpler means. The goat would almost certainly try to get the sacking away with her teeth, and both she and her owner would find the nasty mess unpleasant.

A sharp pocket-knife is generally used for the job, but some people, especially women goat-keepers, prefer a sharp pair of pincers. These work very well and are certainly safer in the hands of a novice or anyone with weak wrists, but the pincers should be kept for this purpose and not used for all sorts of odd jobs, as they are of little use for hoof-trimming unless they are really sharp. A knife does the job more quickly, but you have to be careful, or you may injure the goat or yourself, especially if the horn is hard. The part to be trimmed away is the horn rim, which is the equivalent of our finger-nails and toe-nails. If this gets too long, it bends over the soft pad of the foot and the goat walks on the horn,—just as our own nails bend over, if allowed to get too long. This must be cut back neatly, so that it is on a level with the soft pad and the goat walks on a level foot. The heel should be similarly trimmed. Be sure you get the front point done neatly.

Instructions given for hoof-trimming often sound complicated, but actually it is quite a simple job. Just remember it is the equivalent of cutting back your own nails to a neat shape,—*not* the modern, fashionable claw!—and then look at a hoof, and you will see, at a glance, what wants doing. If you have never done a hoof before, go easily. Do not attempt to cut right back at the first shot, especially if the horn is hard, or you might cut the goat's foot or your own arm. Take off a little at a time, slicing off strips, until you have the whole foot to your taste. When you are used to the job, you can, if you have the horn nice and soft, cut right back first time, especially if you do the hooves regularly and never let them get overgrown. Use the tip of your knife to dislodge any grit or dirt that has got wedged up between the edge of the horn and the soft sole, or the frog, as it is sometimes called.

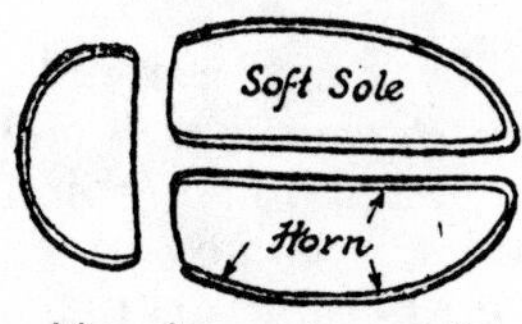

Neatly trimmed, with horn level with frog.

Neglected goats often arrive with their feet in a very bad state. Sometimes the horn overlaps, right across under the foot, so that the goat walks uncomfortably and makes a rattling noise as she goes.

Neglected hoof, showing overgrown horn bent under foot.

This is because she is walking on the hard horn, instead of the soft sole. Sometimes the horn turns up in front, like Dutch "Klompen"! A case like this has to be tackled in instalments, getting the hoof nice and wet and paring off what you can at a time. Hooves should be regularly trimmed from kidhood, and, if you are unlucky enough to buy a goat with badly neglected feet, you may find you never can get them into really nice shape. However, you can, at least, make her comfortable.

If you are unused to this job, or if you have a troublesome goat, it is best to wedge her up against a wall and get someone to hold her steady. Some animals object very much to having their hooves trimmed, even though it is perfectly obvious you are not hurting them, and they will not only kick but will even lie down and sham death! This feigning of death is a favourite trick among goats, and they are excellent actors and do it most realistically. In fact, many a goat-keeper has been greatly distressed, thinking that he, or she, had somehow managed to kill an animal when attending to it, only to hear a reproving bleat and find the "patient" somewhat impatiently waiting for attention!

If a goat does not like what you are doing to her, even though it be perfectly harmless and painless, she may suddenly go limp and slump down, lying stretched out and rigid, with glazed eyes. I have had many a goat that did this whenever she did not want to obey the call to come into her stall. We just left her and took no notice, and eventually she would look about her crossly, get up, and follow on behind. When indoors, a cross goat would refuse to eat,—until she thought nobody was looking,—when she would plunge her head into her manger and make up for lost time! There is no doubt about it that goats have decided and individual characters. Some grumble, if they think you are late with the food, making a very human sort of noise! And some snort and splutter at a food they do

BRITISH SAANEN GOAT
M5 Champion Hartye of Weald, showing the churns she filled when she yielded 479 lbs. of milk in one year. The property of the Misses M. and V. Window Harrison.

BRITISH SAANEN KIDS
ung relatives of the famous RM5 Champion Hartye of Weald. The property of the Misses M. and V. Window Harrison.

[" *Sport and General* " photograph

BRITISH TOGGENBURG GOAT
R4 Breed Champion Northmoor Gazelle, *.
The property of Miss Joan F. Fawcett.

BRITISH TOGGENBURG GOATLING
Northmoor Treasure. The property of Miss Joan F. Fawcett.

not like, quite clearly telling you that you ought to know better than to give them such stuff ! Like children, they give up their tricks when they find they do not work, though they may try them again on any later occasion, if they think they can " pull off something ". And the disgusted and offended look in their eyes, when they find you are not impressed, is too ludicrous for words !

CHAPTER TWELVE

MILKING, AND THE CARE OF MILK

THE gentle art of milking, like a good many other things, looks very easy ; but, when you come to try for yourself, you may find you simply cannot get a drop ! However, do not let this worry you, for it is a knack that comes to you suddenly. After that, it is only a matter of time before you are an expert, swishing your two gallons into the pail in next to no time !

Before you start milking, make yourself comfortable. Squatting on the haunches is not a comfortable position, and, if you feel fidgety, the goat will sense the tension in your attitude and get fidgety as well. You have to approach the job in an easy, confident manner. Use a small stool to sit on, and have a pail for milking. Fortunately, the modern dairy goat is generally tall, and it is not difficult to get a small pail under her. Do not, however, use too high a pail, or you will have the udder partly in the pail, and this worries the goat and makes it difficult for you to use your hands properly. Some people use cans or bowls, but these are very easily upset and a small pail is infinitely preferable.

See that the udder is clean. Rub it over with your hand, to clear away any bits of straw that may adhere, and, if it should be actually soiled, cleanse it with a damp cloth. Washing is very rarely necessary, since the goat's excreta is in dry form. Unless a goat is very ill, with severe diarrhœa, the udder cannot possibly get into the filthy state which is customary with cows. Washing should never be done, unless absolutely essential, as it is liable to result in chills, and a chill is often the forerunner of the dreaded mastitis.

Now edge your goat gently up against the wall of her stall, draw up your stool with one hand and put the pail under with the other, sitting down in the same movement and keeping up as close to the goat as possible. Done in this way, it helps to start you off right. The goat feels you know your job, and she has little room for antics! Grip the pail with your knees and tilt it slightly forwards,—i.e. towards the other side of the goat,—and grasp both teats gently but firmly, resting the lower parts of your arms on the rim of the bucket. Squeeze each teat in turn, making the action brisk *and avoiding any pulling movement.* It is this pulling down and tugging that hinders so many would-be milkers. It is a wrong movement altogether, and the goat objects to it, as well she may. What appears to be a pulling-down movement on the part of the quick, expert milker is an optical illusion, which will be dealt with again later.

Put all thought of pulling downwards out of your mind, and imitate the action of the kid. It takes the teat into its mouth and then stands with its head turned upwards, apparently quite still. Of course it is sucking,—i.e., it is alternately squeezing and partly releasing the teat, by the action of suction between the tongue and the roof of the mouth,—but it quite definitely does not pull the teat down or move its head. This is what you must imitate with your hands. Using your fingers against the palm of the hand,—just as the kid uses its tongue against the roof of the mouth,—press and partially release, repeating the action rapidly and making it as like a suckling kid as possible. Be gentle, firm, and confident. Do not tug, and do not be slow. If the goat gets the idea that you do not know your job, or if you worry her by being slow or hurt her by tugging, she may hold her milk from you, and you will think there is no more to come. This is a bad business, as Nature adjusts the supply to the demand, and, if you do not milk a goat quite dry every time, you will find the supply becomes less and less.

If you are using the right action, the milk should come freely, in good streams, until you have had it nearly all. The streams then become thinner, and you can feel that the udder is about dry. You now have to do the stripping, to make sure you get the very last drops. Incidentally, most of the cream is in the strippings. Some people use the thumb and finger for stripping and some just the first two fingers, but the action is the same. You draw the thumb and finger quickly down each teat, repeating the action, time after time, until the last of the milk has been drawn. Often you can get quite an appreciable amount more, when you think it is all drawn, by bunting the udder with the back of your hand, as the kid does with its head. This is a sort of bumping action, and kids do it so roughly that one often wonders the goat can stand it. The human milker is generally more considerate!

Now, regarding the optical illusion that leads some people to think there is an up-and-down pulling movement in milking, this is partly due to the bunting action just described. As the back of the hand bunts the udder and the fingers slide down the teats, there certainly is the appearance of a tugging movement; but this is not so, as the fingers are sliding down the teat, not clutching it. Again, when milking begins and the squeezing action is used, it does look as though the hands of a rapid milker are going up and down all the time; but this is another illusion, for it is only the lifting of the fingers, as they release their pressure on the teat, that gives this impression.

Once you have got the knack, milking is no trouble at all, unless you happen to have a troublesome goat. Some are very trying, especially when they change hands. Goats that have been milked by a man sometimes think the gentler hands of a woman do not mean business; and those that are used to a woman's hands often think a man too rough for their taste. If a goat is restive at these times, look and see if you are pulling hairs as you milk. This is painful, and, if you find there are long hairs that always get in the way, it is best to cut them off altogether. Again, it may happen that a goat has warts on her udder, and these are best removed by a ligature, when you are drying her off, prior to kidding.

Sore teats, or excessively dry hands on the part of the milker, may make the process painful to the goat. The teats should not be sore, if you are careful to leave her with a properly dried udder; and, if the dryness of your own hands makes milking a slow and painful business, it is permissible to lubricate them with a few drops of milk, so long as your hands are perfectly clean. This is called wet milking, as opposed to dry milking, and it is strongly condemned with regard to dairy cows. This is not surprising, for the results, in the case of cows, are often indescribably filthy. Many a time, I have seen cows come in with absolutely disgustingly dirty udders, to be roughly wiped with a handful of straw and then milked by a farm labourer with equally filthy hands. Afterwards, the palms of his hands and the lower part of the cow's udder would be nice and clean, and the filth would be at the bottom of the pail!

However, goats, fortunately, do not pass their fæces in liquid form, and, consequently, they do not lie down in dirty litter. Their udders are clean, and the hands of the milker should be clean, and lubrication with a few drops of milk certainly makes milking easier and avoids painful friction. But do not make it really wet milking. Mere dampness is all that is needed.

An awkward goat that is determined not to be milked is capable of getting up to all sorts of tricks, and you have to be strong-minded and get her into order. Kicking the pail over is the simplest dodge, and you can easily prevent that by keeping your arms across the bucket, as explained earlier in this chapter, and keeping a wary eye on the legs. Other dodges are trying to sit down, with the udder in the pail, lying down completely, putting the foot in the milk, and trying to stand up on the hind legs. The last-named dodge can be prevented by tying the goat up more closely, before you begin to milk; but, if she really is determined to be troublesome and thinks she can get the better of you, this will make her still more determined. So watch those hind legs! You can keep the near one out of the bucket, by pressing it back with your leg and elbow and hoisting your elbow every time she makes a move with it. She finds this awfully trying, because you can effectively baffle *that* leg; so watch the other, and just tilt the bucket sideways, if she makes a move with the leg.

Lying down, and trying to squat down with the udder in the pail, are much more difficult dodges to combat. Actual lying down, full length, can be prevented by the use of a "guillotine" milking-stall; but there seems to be no way of preventing a goat going lower and lower, and leaning on you more and more heavily, until her udder is firmly in the pail and milking is out of the question. This always strikes me as the most irritating dodge of all! However, if you cannot actually prevent her doing it, you can make it so unprofitable to the goat that she gives it up of her own accord. Tie the goat firmly and closely to her staple, push her up against the wall of her stall, and then push a box underneath her fore-part. Strap or tie her hind legs together, and then take a small receptacle for milking, holding it with one hand and milking with the other. This means you take twice as long milking, but you do get the milk, and the goat cannot do anything about it; so, in due course, she decides it is better to behave herself and let you get the job done more quickly!

She can, of course, still droop herself a little, even with her hind legs strapped, and she will endeavour to do the only other thing left to her,—lean on you more and more heavily. But you must be up to this dodge and keep well away, so that she cannot lean at all. Deprived of her leaning post, unable to bow her legs on account of the strap, and unable to pull away from the staple, she finds her awkwardness is pretty futile. She is very uncomfortable, and yet you are getting that milk! After a few days of this sort of thing, most goats give it up as a bad job. But keep the strap handy, in case it is needed again.

A strap is, also, useful for goats that try kicking and putting their feet in the pail ; but I never recommend using them if you can manage by other means, as goats are liable to become obstinate and try worse tricks. A goat that has had her legs strapped for kicking, or putting her foot in the pail, will probably retaliate with some other dodge, such as rearing up on the hind legs and pawing at the wall. Then you have to tie the head more closely, and one thing leads to another. Goats are intelligent animals and learn by experience ; and, if you just have a lively or fussy one and not a deliberately vicious one, it is generally possible to show her that leg-lifting does not work, by the simple use of the leg and elbow, as explained earlier.

One other dodge is holding the milk. When a goat does this, she humps her back, and you can make her let down the milk by giving her a sharp slap on the back. In due course, they give up the game. Cows which habitually hold their milk are stopped by having a cord knotted round the body, a little in front of the udder,—in fact, in the very place that the back arches, when the milk is held. A stick is then inserted under the rope and given a twist or two, not enough to hurt the animal but just enough to make her let down the milk. However, I have never known such a persistent milk-holder in the caprine species that she had to be roped. Generally, a few slaps on the hump, as it rises, put a stop to the dodge.

I hope all this advice about how to deal with troublesome goats has not given the reader the impression that they are usually difficult creatures to milk, for this is not the case. Many goats are expertly milked by little children of five, or less. The *average* modern dairy goat is quiet, and gentle, and very anxious for you to be fond of her. I have known goats that bleated pathetically, when scolded! In this respect, things are very different from what they were when I first took up goats. In those days, it seemed as though "awkward" was practically every goat's middle name, and guillotine milking-stalls were very common. Now they are rare. In my opinion, this is due to breeding for heavy milk production, for the heavy milker, like the heavy layer, always seems to be more placid and affectionate than the poor producer.

For those who think they ought to have guillotine milking-stalls, I will explain their construction; but do not hasten to make one after your first tussle. There are few really bad-tempered goats among the strains that have been bred for high production; and awkwardness over milking is often a passing phase, due to a change of owner, nervousness, discomfort of a full udder directly after kidding, or just plain puckishness and a desire to "try it on the dog"! Incidentally, with regard to milking a goat soon after it has kidded, it should be noted that she should not be stripped during the first three or four days. Taking every drop of milk, at this time, may result in collapse, due to calcium deficiency, with unconsciousness and death, unless a veterinary surgeon is called in, in time to give an injection.

Guillotines can be fitted to almost any existing milking-stall, by any handy man or woman, in about half an hour. Or they can be attached to milking - platforms. Some people are very fond of the latter, especially if they are tall and find bending a trial; and there is seldom any difficulty in getting the goats to leap up on to the platforms, once they know their use. To make a guillotine, nail a couple of vertical battens down each side of the stall, about on a level with where the neck comes at feeding time. Put each pair of battens about an inch or

The guillotine boards slide down in grooves, and can be removed for cleaning.

so apart, so that they make a groove. You now need several pieces of boarding, about once inch thick. The width is immaterial ; but the wider the boards, the fewer you will need. Cut them to the right length, to run down in these vertical grooves, and carry on until you have, in effect, a little wooden wall a trifle higher than your goat.

Now, with a pencil, mark out an oval where the goat's neck will be, making it big enough to be comfortable but not big enough for the goat to be able to withdraw her head. Starting with a small hole, made with a brace and small bit, cut out this oval with a keyhole saw. The boards should have been so arranged that half of this oval is in the top board and the other half in the one beneath ; and, needless to say, the longer side of the oval should be the depth, and the narrower should be the width. All you need now is a couple of hooks and eyes, to keep the top board secured when the goat is in position. They leap up quite willingly, when they get to know that there is food on the other side of the small, wooden wall ; and, as soon as a head is pushed over, you bring the top half of the oval into position and fasten the hooks and eyes.

When these guillotines are attached to milking-platforms (and, also, with some types of stalls), the board in which the top half of the oval has been cut is hinged on one side and is brought across (instead of being brought down), and is fastened on the other side with a hook and eye or hasp and staple. This does away with the necessity for having two battens on either side, as all the boards but the top one can be nailed into place, unless this makes cleaning difficult. In the case of a milking-platform which is movable, this latter difficulty would not arise. The use of a hinge on one side often makes a stronger job of it ; and it is preferable, if the wood you are using is not well seasoned. Green wood that warps after cutting is sometimes troublesome in the grooves.

Some people give their goats their concentrates

to eat while they are being milked, and this is quite a good plan if they are quiet and well behaved and you have little time to spare. It is, also, a good way of soothing nervous goats and getting them used to being milked. But, as a general thing, I prefer the goats to stand, quietly cudding, while they are milked. A cudding goat is a quiet and contented goat. They do not cud unless they feel contented. The mental state induced by cudding is placid and easy-going—a state in which the goat will let you have her milk without trouble, because her own thoughts are fully occupied with a very pleasant process at the other end! This is how it should be.

A goat that must be allowed to eat while she is milked is anything but placid. The food excites her, and she is pushing and poking at it, and jerking her body about in a tense way that does not encourage the flow of milk. Also, if she finishes eating before you have finished milking, she is likely to get restive and make a fuss. I always teach my goats that they will get their concentrates after they have been milked, so long as they behave. If they waste my time by their antics, they do not get any concentrates. It does not hurt any goat to make do with hay. It only annoys them—which is a good thing, in those particular circumstances! The milk yield may drop a bit, but it is worth it, to enforce discipline; and it soon rises again, when you have won the battle.

Goats, as I have said before, are very intelligent animals, and they soon learn that concentrates come *after* milking to *good* goats. Then they stand and cud peacefully while you milk, and you know you have come to a definite understanding. Sometimes, you will find a certain nanny tries to pay you back, by telling you what she thinks of you. After you have milked her in peace and quietness, you take her her concentrates, and she bangs her head into them, whacks your arm out of the way, and makes rude noises at you! But she soon gets over it. She is just peeved, because she has had to give in to you, and she cannot do it with a good grace !

Strict cleanliness must be observed in connection with milking, especially if you want to make butter and cheese. Unsatisfactory results in the making of these delicacies are often due to nothing else but careless handling of the milk, which causes undesirable bacteria to develop. Directly after milking, the milk should be taken away from the goat-house and strained. This can be done with a wire strainer of very fine mesh or a piece of clean muslin. Or the proper dairy filters can be used —in which case you have a fresh disc each time, and this saves a certain amount of washing and sterilising. Straining is desirable, even if you are going to use the milk at once and are quite sure it is clean, as straining removes hairs.

The milk should be cooled as quickly as possible after straining, if necessary by standing the utensil containing the milk inside a larger one containing the coldest water you can obtain. The quicker it cools, the better it will keep. It should then be covered with muslin, to keep out dust and flies, and put in a cool, airy place.

All utensils used should then be washed and sterilised immediately. First, they should be rinsed with cold water ; then they should be washed thoroughly with hot, suddy water, in the usual way. Finally, they should be rinsed again in cold water and then sterilised with boiling water. Dry with spotlessly clean cloths, and hang up in a clean and airy place, and you should have no trouble through lack of cleanliness. It is often said that butter-making begins in the milking shed, for it is only after it has been drawn that the bacteria can gain access to it. Remember this, if you are ever tempted to skip the careful washing and sterilising. Some bacteria are, of course, good. Without them, we could make neither butter nor cheese. But undesirable bacteria, which thrive in uncleanly conditions, may give your butter a very unpleasant flavour, as well as making it difficult to churn.

CHAPTER THIRTEEN

BUTTER AND CHEESE FROM GOATS' MILK

MANY people have the idea that making butter and cheese are tedious and laborious jobs, but this is not necessarily the case. It *can* be tedious, and it *can* be hard. Remembering my own early days, when I made butter from cows' milk, with an " end-over-end " farmhouse churn, I can understand the feelings of readers who want to skip this chapter. But please don't! Butter-making with a modern, glass table-churn is no more trouble than making a cake and, to my mind, a good deal more interesting. There is a fascination in watching the cream swirling round and seeing the butter-grains form.

You do not " have to go on churning for hours and hours ". At least, if you do, there is something wrong. If your cream is properly ripe, is at the right temperature for churning, and is free from undesirable bacteria, the butter should come in about half an hour. If it takes more than forty minutes, there is something wrong ; and, if it takes less than twenty minutes, there is something wrong. Butter sometimes comes in about five minutes, much to the joy of the novice ; but this is due to some fault, such as too high a temperature, and the butter thus made will be greasy and will not keep well. It is not difficult to make good butter. It only means paying strict attention to certain important details, which you do without thinking after you have made a few lots. It is not hard work, and it does not take long.—so now, perhaps, you will read this chapter! Remember that, even with strict rationing in force (as it is, at the time of writing), you can have all the goats'-milk butter and cheese you like and can sell any surplus. It is unrationed and free of all control and regulations, except that the butter is price-controlled. If any local know-all tells you otherwise (as they have even been known to do at the Food Offices), you can tell them to study their own rules!

Before you learn how to make butter, you must, of course, know how to treat the cream while you are holding it. You can, of course, use a separator, to take the cream from the milk ; but most goat-keepers operate on too small a scale to justify the purchase of a separator and just skim off the cream with a large spoon or a skimmer. The advantage of a skimmer —which is like an almost flat, rimless plate, with holes in the centre and a handle at the side.—is that it makes it easier to skim off the cream without taking any milk.

As soon as the milk has been strained and cooled, pour it out into shallow pans or dishes and leave to stand for anything from twenty-four to thirty-six hours. Be sure the dishes stand quite level. Then skim it, being careful not to take milk with the cream, and put the cream into a large bowl or crock. To each breakfast cup full of cream, add a small teaspoonful of saltpetre and a large teaspoonful of salt. Each time you add more cream, stir well, adding a little more salt as you do so. After about four days, the cream should be ready for churning, though this depends a good deal on the temperature of the room where the cream has been kept, and, in winter, churning is generally only done once a week. Nearly six ounces of butter should be obtained per gallon of milk, and gravity-skimmed goats' milk (i.e., milk skimmed by hand, after the cream has risen by the force of gravity, which keeps the heavier milk below), still contains a good deal of cream and is as good as much unskimmed cows' milk. This is because the cream globules in goats' milk are much smaller than is the case with cows' milk, and they do not rise so readily. In addition to this, you have, also, the buttermilk, which is excellent for making scones, for drinking, for the complexion, and for use in cheese-making. So butter-making is well worth while.

Do not bother about having elaborate equipment

for your first attempts at butter-making. It will be soon enough to think about this, when you are making an appreciable amount regularly. Actually, it is possible to make butter, just by shaking up the cream in a jam jar, or by whisking it up with an ordinary egg-whisk in a pudding bowl. However, it certainly makes it easier if you have a little churn, and the glass kind that you stand on the table is quite large enough for the average family. They can be obtained from agricultural ironmongers and dairy supply firms, and they are complete with a paddle attached to the lid, the paddle being worked by turning a little wheel round and round—very much like an egg-whisk.

In addition to this, you should have a dairy thermometer, for correct temperature is very important. " Scotch hands " are very convenient for patting up the butter, but you can manage very well with two pieces of smooth, clean wood. You need, also, the pastry board, (or any other piece of smooth, clean board), and a clean, soft cloth, for mopping up the moisture you pat from the butter. If you do not like to see white butter, you should have a bottle of annatto (obtainable from agricultural ironmongers and chemists), but this does not affect the taste and so is not essential. These few things are all you need, and it seems to me a much simpler business than making cakes and pastry. There is no weighing, mixing, beating, kneading, rolling, dashing about for this and that, and squinting into the oven to see what is happening! You just turn the little handle round and round, while you think about something else!

Strict cleanliness, correct temperature, and ripeness without staleness of the cream are the secrets of making good butter, with a nice flavour and texture and of good keeping quality. Do not add any fresh cream within twelve hours of churning. Strain the cream into the churn through a piece of scrim, or other open-meshed cloth, and add annatto, if you wish. As goats' milk has no natural colouring matter, your butter will be white without it, but it will taste just as good. How much to add is a matter of choice. Try a little at a time, until you get the colour you fancy. Now add hot water, till you have the cream at the correct temperature, which should be from 55 to 60 degrees Fahrenheit; but do not let the churn be more than a third full.

You now begin churning, turning the handle slowly at first, and ventilating, from time to time, to allow the gas to escape. When gas no longer forms, you can work up a good speed, and soon you should see the pin-head particles of butter coming. Now add water at four degrees below churning temperature, to protect the grains and prevent them lumping together. Continue churning, adding more water if required, until the particles are the size of wheat grains. It is very necessary to add this water, as, if you churn the butter into a lump in the buttermilk, it will develop a very nasty flavour and will not keep well. The cooler water keeps the grains apart.

When the particles are the size of wheat grains, you have really made your butter. All you have to do now is to strain off the buttermilk, wash the grains, salt them, and make up your pats. The washing water should be poured in at four degrees below the temperature of the buttermilk. The quantity is immaterial. Give a few turns, to wash the buttermilk from the grains. Now pour it off and add cold, salted water, and leave for fifteen minutes. The amount of salt is a matter of taste. Try six ounces to a quart of water. When it has taken up enough salt—i.e. in about fifteen minutes—pour off the water and put the butter on the board and pat it into shape. Do not work it too much. You only pat it about to get the water out, and this should be mopped up with a bit of clean muslin, or soft, old teacloth, as you go along. The butter is now ready for use.

If your first attempts are not as successful as you could wish, do not be discouraged but try to find out what was wrong. If the butter has an unpleasant flavour, it is probably because you did not wash out the buttermilk thoroughly. If it is lacking in flavour, no doubt your cream was not sufficiently soured. It is no good trying to remedy this, however, by keeping your cream longer before churning, as this would only make it stale, and stale cream does not make good butter. If you want a fuller-flavoured butter, you should add a little of the sour cream, held back from the previous lot, to the fresh cream you put in the crock. This starts the souring, or ripening, process. If you are making a lot of butter every day, you will, of course, use a commercial starter, which is a culture obtained from dairy institutes and sours the cream in about twelve hours.

If the butter breaks—i.e. forms into particles—too soon, the temperature of the cream was too high at the time of churning. But, in Great Britain, it is far more often the case that the temperature is too low, especially in the winter months, and often the cream will not ripen at all in the creamery. It just gets old and goes stale. This means hard, laborious churning and poor quality butter; and sometimes it is even impossible to get butter at all. If you find this happens in the winter, the only thing to do is to keep your cream in a warmer place while it is ripening. Your best guide is the state of your cream after four days. If it is not ripe enough after four days to give butter in about half an hour, the temperature of your creamery is too low. If it sours too quickly, the temperature is too high.

CHEESE

A number of delicious soft cheeses can be made from goats' milk, also hard, pressed cheese and Stilton cheese. The two latter are, of course, more trouble to make and take longer to ripen, but soft cheeses are ready for eating in a matter of hours and are generally preferred by children and those who do not care for a full-flavoured cheese.

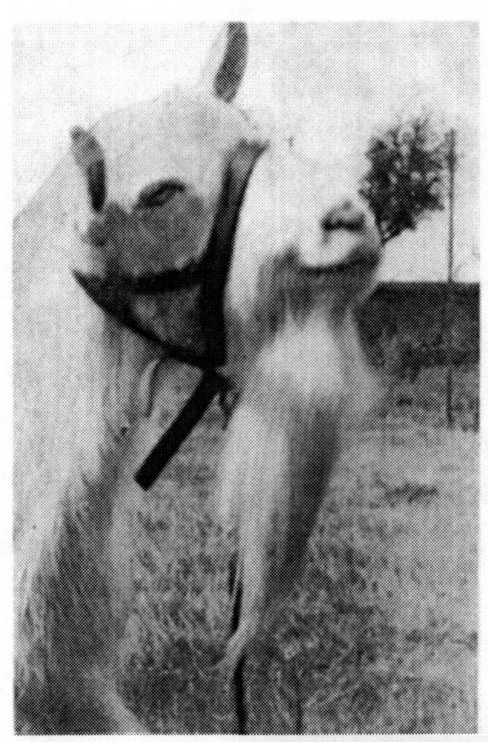

FINE HEAD STUDY OF BRITISH SAANEN MALE
Hillam Rufus. The property of Ernest S. Smith.

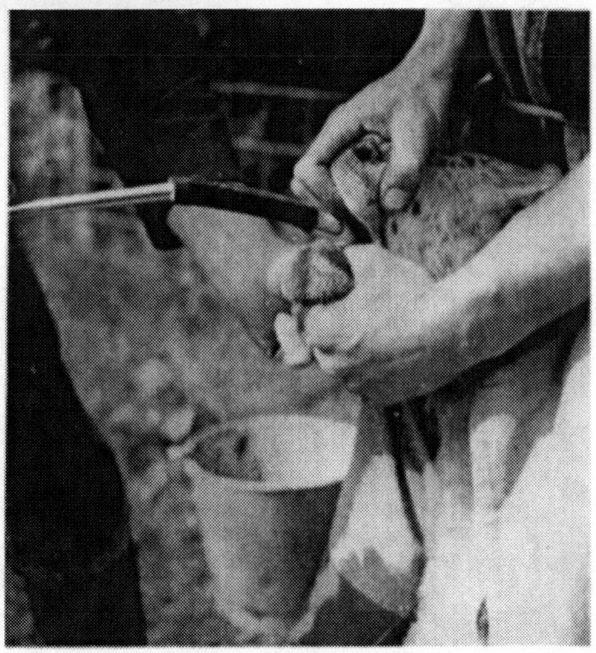

PILL CARRIER IN USE
See page 169.

[Photograph by courtesy of Messrs. Boots, Ltd.

GOATLINGS

Promising young members of the Weald Herd. The property of the Misses M. and V. Window Harrison.

[" *Sport and General* " photograph.

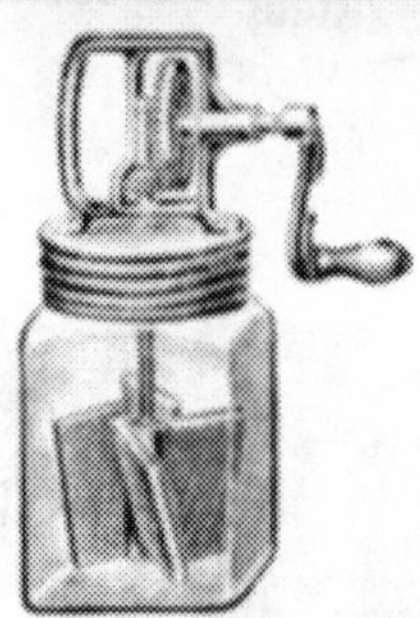

↑ GLASS TABLE CHURN

The property of the Dairy Supply Co., Ltd. This little glass butter churn is ideal for the goat keeper, as it is quite a small thing and stands on the table. It is available in various sizes, from 1 pint to 6 pints. See page 140.

JACKETTED VAT →

The property of the Dairy Supply Co., Ltd. This vat is the very thing for the small-scale cheesemaker, as it is available in sizes from 15 to 25 gallons. It saves much trouble, when you have to raise the temperature of the milk or curd by degrees. See pages 148 and 149.

A universal favourite is Bondon cheese, and it is convenient to make this when you have been making butter, as part buttermilk can be used. Mix two pints of buttermilk or sour milk with every one and a half gallons of whole milk and heat to 70 degrees Fahrenheit. Add one cubic centimetre of rennet to every gallon of the mixture and stir carefully. Leave to set for two hours, and then cut the curd and ladle it into cloths for draining. When it has drained, it should be transferred to fresh cloths, (or placed in little moulds lined with cheesecloth and placed on straw mats to drain), and left for from one to two days. Just how long it should be left depends on your own individual taste, but it is ready for eating as soon as it has finished draining. This amount of milk should make twelve cheeses of the usual Bondon size.

Although it is always simpler to do a job when you have the proper appliances, it is not essential that you should have moulds and straw mats for making these little soft cheeses and others of similar nature. I have used cake tins with removable bottoms, lining them with muslin and placing them on a clean teacloth. All that is needed is that the moisture should drain away from the curd. I have, also, made soft cheese very quickly by adding rennet to milk, as though making a junket, sliding it into a colander when set, and leaving the colander over a pudding basin, to drain all night. By the following morning, you have a pleasant soft cheese that is popular with children who do not normally like cheese; and, by evening, it has a good flavour.

This method is my own and, of course, quite unorthodox. Nevertheless, it shows that you need not be deterred by lack of appliances from making surplus or sour milk into cheese for the table. The chief essentials to success in cheese-making are cleanliness of the milk, correct temperature, and careful measurement of the rennet. About these things you must be very strict. The rennet is measured by cubic centimetres, (c.cms.), and it is worth making a note of the fact that fifteen drops equal one c.cm., and one teaspoonful equals 4 c.cms.

So long as you remember to observe these three rules carefully, it does not matter a great deal if you have not the proper appliances; though it is more convenient to have them, if you make cheese for sale. For home use, the actual size of the cheese, the degree of ripeness, colour, and flavouring are matters for individual taste and leave ample scope for experiment. It is usual to add a little salt to the curd, when cutting it. Some people use pepper as well, and some like celery salt. If you want your cheese to be coloured, add a few drops of annatto to the milk, before you add the rennet. The cutting of the curd is done with a long knife or ladle, but it can be managed with an enamelled egg-slice.

Soft cheeses can be made from all cream, cream and milk, whole milk and buttermilk, or whole milk and sour milk, so there are quite a variety from which to choose. Cream cheese can be made with or without rennet. To make cream cheese without rennet, you should take very thick cream and cool to 65 degrees Fahrenheit in three hours. Put it into cloths and hang to drain in a fairly warm place in a draught. At intervals of six hours, take it down and turn it, sides to middle, and scrape down the sides of the cloth. This assists drainage. Salt it to taste when nearly set and mould while still sweet.

To make cream cheese with rennet, take equal quantities of cream and new milk and mix thoroughly. Raise the temperature of the mixture to 80 degrees Fahrenheit and add 1 c.cm. of rennet to every gallon. When the curd has set, cut it into squares and ladle into cloths to drain. When cutting curd into squares, it is usual to cut it into pieces about the size of walnuts, unless other directions are given. The cross cuts are made both ways, and then the ladle or skimmer can be used to make the horizontal cuts.

Gervais cheese is made of two-thirds whole milk and one-third cream. Mix well and add one or two drops of starter. Raise the mixture to a temperature of 78 degrees Fahrenheit and add $\frac{3}{4}$ c.cm. of rennet to every gallon. Stir well, and leave for two or three hours. Cut the curd into 2 in. squares and ladle into cloths for draining. Three quarts of this mixture should make twelve cheeses.

The longer a cheese is in the making, the more flavour it has, and those who find some soft cheeses unsatisfying should try Coulommier. This can be eaten after three days, but it is better if left to ripen for three weeks. So, as regards time, it is a sort of half-way house between soft and hard cheeses, as the latter take from three weeks to three months to ripen, according to variety.

To make two Coulommier cheeses, take one gallon of new milk and add a small teaspoonful of starter. Heat the milk to 86 degrees Fahrenheit, and add 1 c.cm. of rennet, diluted with five parts of water. Deep-stir for three minutes and then top-stir till it coagulates, and leave it to set. Since these cheeses are usually left to ripen, it is best to have the proper moulds and straw mats, and these should always be scalded before use. Place a scalded mat on a board,—the pastryboard will do, if clean and scalded,—put a mould on the straw mat, and ladle the curd into the mould. It is usual to take out a nice-shaped piece first with the skimmer and keep this for putting on top, when all the curd has been ladled into the mould.

By the following day, the curd should have sunk to the top of the bottom portion of the mould. Scald another straw mat and board, put the mat over the top of the mould and turn the cheese over, leaving it standing on the fresh mat until the next day. By the third day, the cheese should be firm enough to stand without the mould. Dry salt should be rubbed over the surface, and it should be stood on the board and turned daily until it is wanted. It will be ready for eating in a few days but is better flavoured if left to ripen for three weeks.

Hard, pressed cheese is more trouble to make, but it is well worth while, and it is not essential to have much special equipment. What you must have are a reliable thermometer and a suitable mould. If you are making regular quantities of cheese for sale, it is best to have a jacketed vat for heating the milk and, later, the curd; but, for making small quantities for home use, you can manage very well by placing one receptacle inside another one containing hot water. Again, a cheese press is a help; but, for home use, you can manage with the motor-car jack.

To make one cheese of about six and a half pounds, you need six gallons of milk. It ought to be all milk of the same day's milking, but home cheese-makers, who have not the amount to spare in one day, often save it for a few days. This is all right, so long as the milk is clean and kept in a cool place. The quality of the cheese may vary, but it should all be good cheese. The milk should be strained into a tub, or other suitable receptacle, and raised to a temperature of 90 degrees Fahrenheit. This is where a jacketed vat saves a lot of trouble; but you can manage, by standing the receptacle containing the milk inside another one containing hot water. Take 5 c.cms. of rennet, diluted with six times their volume of water, add to the milk, and stir thoroughly for three or four minutes. Then top-stir until the milk begins to coagulate. Cover the receptacle and leave it until the curd is firm and springy, when it will be ready for cutting. This will probably be about half an hour from the commencement of coagulation. The curd should now be cut into cubes of half an inch or less, using a long knife that will reach to the bottom of the receptacle for the vertical cuts and a skimmer, (or, failing that, an egg-slice), for the horizontal cuts. Proper curd-cutting knives can be bought from dairy supply firms and sometimes from agricultural ironmongers. After cutting, stir the curd gently with the hands.

The curd must now be scalded, and this, again, is

an occasion where a jacketed vat is a great help. However, improvisation is a very good thing, when you are trying your hand at a new job. It will be soon enough to invest in more convenient equipment, when you have made up your mind to carry on with it regularly. The temperature of the curd must be raised to 106 degrees Fahrenheit, but it must not be raised faster than one degree in three minutes. This can either be done, as with the milk, by placing the receptacle inside a larger one containing hot water, or by straining off some of the whey, heating it, and returning it to the receptacle. If the latter method is adopted, the heating must be done in several stages, raising the temperature of the curd not more than five degrees at a time. Admittedly, this scalding of the curd is a delicate process, and it is essential that you should have a reliable dairy thermometer and be very accurate about temperature. Otherwise, you may spoil your cheese. Never raise the temperature of the whey above 120 degrees Fahrenheit.

Whichever system you adopt, keep stirring until the curd is firm and the particles, when pressed together in the hand, do not crush. Strain off the whey, and spread the curd evenly over the bottom of the vat, adding salt at the rate of three ounces to every six gallons of milk.

For this amount of milk, you will need a mould six-and-a-half inches in diameter and the same depth. It should be perforated all round and have two "followers",—firstly, a tin disk, and then a wooden one. The various moulds for different types of cheeses are obtainable from dairy supply firms and agricultural ironmongers. Line the mould with cheese-cloth and pack the curd in carefully. Never be rough and heavy-handed, when you are handling curd. Fold in the ends of the cloth and put in the tin follower and then the wooden one, and put the cheese to press with 2 cwts. pressure.

If you have no cheese-press, it is a good idea to start with bricks, allowing ten pounds weight to every pound of curd. After six hours, substitute the car jack, and screw up, from time to time, as required.

Turn the cheese the same evening, and return it to the press. Next morning, take it out and sew it in a calico bandage, and put it back to press for one hour. (For home use, many people dispense with the bandaging. The cheese may grow mouldy on the outside, if not bandaged, but this does not matter much.) Place it in a fairly dry and draughty place for two days, and then put it on a shelf anywhere where there is a reasonable amount of air. Turn the cheese over from time to time. It can be eaten after three weeks, but it is much better if left to ripen for about three months.

When you have made hard cheese successfully, you may like to embark on that delicacy, Stilton cheese. Take two gallons of milk and strain it into the cheese vat, or whatever other receptacle you may be using. Enamelled pails can be used, so long as they are of good quality enamel and free from chips. Heat the milk slowly to 85 degrees Fahrenheit, and add 30 mms. of rennet, diluted with five times their volume of cold water. Stir gently for five minutes, to mix thoroughly, and then top-stir for a few minutes, until coagulation begins. Take care not to overstir.

In about three-quarters to one hour, the curd will be ready for ladling. Spread a coarse, damp cheese-cloth over a large bowl, with the ends hanging over the sides, and ladle the curd, in thin slices, into the middle of the cloth. Fold the ends over, and leave the bundle in the whey until 0.125% acidity has developed. This will take about two hours. For measuring acidity, an acidimeter is needed. The corners of the cloth should now be taken and tied across diagonally, leaving the bundle fairly loose, and some of the whey should be poured off and the bundle left for another three hours. Pour off whey again and tighten the bundle. After another two hours, pour off the rest of the whey and turn the bundle over on to the knot, to facilitate drainage. The whey should now show 0.17% acidity. When 0.2% acidity has developed, the curd is ready to turn out and salt. This state may be reached the same day, but it may be necessary to leave the curd overnight, if the weather is cold.

Turn the curd out of the cloth and break it up into pieces about the size of walnuts. Add salt at the rate of one ounce to every three pounds of curd and mix thoroughly, but handle the curd carefully. When the salt has dissolved, the curd is ready to mould, and this should be done in a certain way, as it helps to give the desired texture in a Stilton cheese and forms a foundation for the coat.

The moulds should be four-and-a-half inches in diameter and six inches high, with holes round the sides for drainage. Place a piece of calico on a board and a mould on the piece of calico, and then begin to place the curd in the mould. Make a layer of some of the smaller pieces and press them lightly with the hand. Then put some larger pieces in the centre, with smaller pieces round the sides. When the mould is full, leave it for an hour ; then turn on to a fresh board and a fresh calico square. Repeat this process of turning on to fresh boards and squares the next morning, and continue to do so, every day, for about a week. By this time, the cheese should be loose in the mould and sufficiently firm to be slipped out of the mould and to stand alone. It should have a smell of ripe pears, and the sides should be greasy.

The surface of the cheese should be scraped and all crevices filled with the scrapings. This helps to form the characteristic coat. After scraping, pin a one-and-a-half-inch calico bandage tightly round the cheese, replace it in the mould, and leave till the following morning. Then scrape the cheese again lightly, preferably with a hot knife, bandage it, and put it on a board without a mould. It is essential that the scraping should be done lightly, or a thick, heavy coat will be formed. Turn the cheese daily, putting on dry bandages, and continue this process until a coat begins to form and dry patches are found on the bandages. When a fine, white mould is seen, bandages are no longer needed ; but the cheese must still be turned daily and put on to clean boards and cloths.

The ripening process takes from two to three months, according to size of cheese. Blueing is encouraged by piercing the cheese with a sharp, clean knitting needle. This admits air and helps the growth of the characteristic mould. The needle should be pushed into the sides of the cheese in about twenty different places, to a depth of two inches. Just how much piercing is needed depends, of course, on the size of the cheese. From this amount of milk, you should have a cheese weighing about one pound ; and it is not wise to attempt to deal with smaller quantities, if a good cheese is desired.

CHAPTER FOURTEEN

HOW TO TREAT A SICK GOAT

GOATS are, by nature, very healthy and hardy animals, and it is rare to see a sick one. So long as you keep them in a sensible manner, you should not be troubled with anything in the nature of a serious illness. However, there are times when a knowledge of first aid and simple remedies is useful, and it is wise to keep certain things by you, for use in any such emergency. Treatment is sometimes needed in a hurry, especially in the case of stomach upsets in young kids. The complaint itself is not what one would call serious, but little kids cannot endure much severe pain, and you may easily lose one, just through not having the right thing at hand.

I do not advise anyone to try to treat a really sick goat without professional advice. If you think there is something radically wrong, or if a goat has an abnormally high temperature, you should send for the vet. and follow his instructions. It is not fair to call in a vet. when you have lost valuable time in trying to deal with what is beyond you. You may have done real harm, in addition to losing time, and the recuperative powers of the goat may have dwindled to the point where she cannot respond to his treatment. There is no shame in calling in a vet., if you have a sick animal. At the same time, a knowledge of simple remedies, and their uses and the correct dosage, will often enable you to relieve pain and save you a bill. It may, also, help you to carry on with whatever the vet. recommends by telephone, if he cannot get out to you immediately.

Here is a list of things the wise goat-keeper should have by him:—

Linseed Oil.
Castor Oil.
Essence of Peppermint.
Chlorodyne.

- Epsom Salts.
- Prepared Chalk.
- Boracic Powder.
- Carbolised Vaseline.
- Essence of Ginger.
- Collodion.
- Insect Powder.
- Disbudding Stick.
- Disinfectant.

These are all useful things that keep well. In addition to these, many people like to keep a bottle of a proprietary udder drench in the medicine chest, and this is a very wise precaution. There are several old tried and trusted favourites used for cattle, and they are equally suitable for goats. Generally the dosage for goats is given ; if it is not, you can give the dose recommended for sheep.

Worm medicines are best not kept in stock. They should be purchased freshly, as required. There are a number of proprietary makes, and some seem to work better than others. When you find a make that pleases you, stick to it. Another proprietary medicine that may well be kept in stock is a soothing mixture that can be given for internal pain and chills. Again, there are various makes, and these are advertised in the agricultural journals by old-established firms.

When purchasing insect powder, it is best to choose one that contains D.D.T. The disinfectant chosen should be a safe and gentle one for bathing such things as sore teats. But you may, also, keep a can of something stronger and cheaper, for general scrubbing and cleaning.

It is easy enough to tell when a goat is feeling out of sorts. She may not make a sound. In fact, goats often make no outcry when very ill indeed. But the first thing you notice is that the coat is up on end and the sufferer looks miserable. If suffering internal pain, she may arch her back and hang her head. If very bad, she may lie down and be unwilling, or unable, to get up again. Her body may be hard and blown up, and she may groan, from time to time. When you have gained a little experience, you should be able to form a good opinion as to whether the trouble is a passing upset, such as we treat ourselves for, or whether it is the beginning of something more serious. If in doubt, take the animal's temperature. This is taken with an ordinary clinical thermometer, such as we use for ourselves, and it is placed in the rectum. The normal temperature of a goat is 103 degrees Fahrenheit, and, if you find a sick goat has a temperature more than a degree or so above or below this, you should get a veterinary surgeon without delay. A high temperature means fever, but a sub-normal temperature generally means that the goat is very near death.

If you find a goat looking sick, the first thing to do is to rug her. This, alone, often effects an improvement, just as we, ourselves, feel rather better, whether we have a chill or a tummy-ache, when we wrap something warm round us. Take a piece of an old rug or blanket, or even a folded sack, put it across the goat's back, and,

with strings tied to the four corners, secure it across her chest and across her hind-quarters, behind the legs. If she insists on lying down, bank a large armful of straw, tied into a truss, behind her, so that she remains in a comfortable, sitting posture, which is the usual equivalent for lying down in goats. It is a bad thing if a sick goat becomes so weak that she is prostrate. You now have to consider whether your patient has digestive trouble or a chill,—the symptoms of which are much the same as in ourselves,—or whether it is something more serious. In any case, you will have done no harm by rugging her and making her comfortable. Here are details of simple treatment for some of the more common complaints, given in alphabetical order:—

Blood in the Milk. This is a complaint needing treatment, for the milk, though wholesome, looks revolting and is only fit for livestock. The goat, however, is not sick and does not need rugging or supporting. The cause is thought to be an excess of rich foods; so proteins should be cut down and the goat fed mainly on greenstuff and hay, until the trouble is righted. Give the sufferer a dose of Epsom salts, (one tablespoonful to a quarter of a pint of water), and let her have a bran mash every day or two. Directions for making a bran mash will be found in Chapter Seven, on Breeding.

Colds or Chills. These are usually due to direct draughts across the body, or to germs brought in on a fresh goat, for goats seldom suffer from actual lowness of temperature. Rug the goat and keep her in a warm, but not fuggy, building. Give her a dose of Epsom salts and follow with doses of chlorodyne, (20-30 drops, in a little warm water), morning and evening. Alternatively, she may be given quinine tablets or aspirins, as for humans, or she may be treated with one of the proprietary mixtures for chills already mentioned. If the condition does not improve quickly, and if the breathing is difficult, a vet. should be called in without delay.

Colic. This is a complaint that affects kids more often than adults, especially bottle-fed kids in the summer months, when it is difficult to keep the milk fresh. Kids scream and throw themselves about, when in pain, and they must be relieved as quickly as possible. First give a tablespoonful of linseed oil, to clear out the offending material, and then give a soothing dose, such as a few drops of ginger or peppermint in a little warm water, or a dose of one of the proprietary mixtures already mentioned. I do not advise giving chlorodyne to young kids. Massage the stomach and sides, if the kid is in great pain. This condition often comes on very suddenly and clears up quickly, if prompt treatment is given.

Constipation. If a goat goes off her food with no apparent signs of actual illness, it is probable that she is constipated. The droppings of a healthy goat should be firm but moist. If they indicate constipation, you should overhaul your management, for this complaint is generally due to overfeeding with concentrates and lack of greenstuff and the necessary bulk foods. It may, also, be due to lack of exercise or insufficient drinking water. In kids, it often occurs at weaning time, owing to the change of diet. If it is a definite case of constipation due to some recent happening, such as change of diet, the goat should be given four ounces of linseed oil and four ounces of castor oil, shaken up together in a bottle. A kid can be given half or quarter doses, according to age and development. If, however, the constipation is a more or less continual trouble, due to some fault in management, remedy the fault at once and give the sufferer two ounces of linseed oil, every other day, for a week or a fortnight, according to the obstinacy of the condition. When you have her right, give her bran mashes two or three times a week, for a few weeks. After this, she should keep right, if you do not let the same fault occur again. Much eating of acorns or chestnuts often results in obstinate constipation.

Diarrhœa. This complaint is generally known as Scours among livestock breeders. It is frequent in the Spring, when the stock first go out to graze, after having been kept in for some months. If it is not unduly severe, it will clear up naturally, as the system becomes used to the change of diet; but it will be eased if you see all goats pack up well with hay, before going out to graze. Any sudden change of diet is liable to cause scours, and it is particularly common among newly-weaned kids. Every effort should be made to introduce changes gradually, both with kids and adults, so that scours may be avoided. Diarrhœa may, also, be due to worms and is, in fact, the accompaniment of many complaints. If it is a symptom of some other trouble, it will clear up as the trouble goes ; but, if the diarrhœa is itself the complaint, then the cause must be sought and the diarrhœa checked.

The checking must not be done suddenly. First, give the sufferer four ounces of linseed oil and two ounces of castor oil, shaken up together, to clear out the irritant material. Then follow this, about three hours later, with a tablespoonful of prepared chalk. If this is not sufficient, follow, a few hours later, with a little chlorodyne in warm water. From fifteen to thirty drops should be given, depending on the age of the goat and the severity of the attack. Or a dose of one of the proprietary pain relievers for cattle may be given.

Diarrhœa in kids needs gentler treatment. First of all, try giving the white of an egg, beaten up with a tablespoonful of water, instead of one of the milk feeds. Give smaller feeds, so as to tax the digestion as little as possible. If this is insufficient, follow it with a teaspoonful of prepared chalk, given in a small quantity of milk. If the kid is eating solid food, it may be given arrowroot biscuits, which often prove beneficial in such a case. If the diarrhœa will not clear up by these simple means, castor oil must be given, the dose varying from a dessertspoonful upwards, according to the age of the goat. A dessertspoonful is a suitable dose for a well-grown kid of six weeks of age. Always follow this, within a few hours, with prepared chalk, as an astringent after turning out the irritant material.

Dysentery. Neglected diarrhœa often turns to dysentery, which is similar to diarrhœa but often contains blood and has a characteristic and very objectionable odour. When it gets to this stage, it shows that harm has been done and that there is inflammation of the bowels. Dysentery may, also, be caused by the animal eating food in a bad condition ; or the germ may be picked up from other animals. Unlike a goat with simple diarrhœa, the sufferer from dysentery often refuses to eat, and this makes it more difficult to pull it round and keep up its strength. The patient will feel the cold, and she should be warmly rugged, kept in a warm, but not fuggy, building, and have a good armful of straw behind her back. Dose, first of all, as for diarrhœa, with linseed oil and castor oil. Follow this with the prepared chalk, in three hours. At bedtime, give from twenty to thirty drops of chlorodyne in warm water, or one of the proprietary cordials sold for this complaint. Continue with the chlorodyne or the cordial, every morning and evening, until a few days after the trouble has ceased.

Foot-and-Mouth Disease. Goats can contract this disease, though it is rare, and cases have been known where goats have been in contact with cattle suffering from it and yet have not fallen victims themselves. Actually, the disease would not be difficult to treat, but it is never attempted, since the danger of infecting other livestock is so great. It is a notifiable disease,—which means you have to notify the police, if you find a case of it among your stock,—and all sufferers and their contacts are destroyed. However, the goat-keeper need not worry about foot-and-mouth disease, since it is practically unheard of among goats. It is only mentioned here because it is illegal to take stock on the roads, or send them in and out of your area, at times when there are foot-and-mouth restrictions. These periods do not, as a rule, last long, and one just has to wait until the " all-clear " is given.

Foot Trouble. This may only mean that the hooves need trimming, or that a thorn or small stone has become wedged between the horn and the soft part of the foot. Foreign bodies of this sort are easily removed, if attended to at once; but do not let a goat go lame for some time without attention, or you may find there is something deeply embedded, so that you have to poultice the foot. Goats do not like this sort of thing and make your work harder by trying to remove bandages and poultice with their teeth. Directions for trimming hooves are given in Chapter Eleven, on General Management; and this matter should never be neglected, or Foot-rot may develop, and this is a very unpleasant complaint to have to treat. This is definitely one of those cases where prevention is better than cure. When the hooves are badly neglected, the horny part grows so long that it folds under the foot and overlaps. Grit gets under the flaps and presses into the soft part, or frog; and eventually this part cracks and becomes very sore and painful, so that the goat tries to avoid walking. If nothing is done about it, the inflammation reaches the "quick", and there may, also, be ulceration. The smell is revolting. However, if you have a goat in this state, you must take it in hand. First pare the horn, being as careful as you can about it. Wash the foot thoroughly, with warm water and disinfectant, and then dress the affected parts with a solution of one part of carbolic acid to twenty parts of glycerine. This mixture should be made up by a chemist. Crude carbolic disinfectant is *not* the right thing. Apply linseed poultices for a few days, until the parts are clean and ready for healing. Then wrap the foot in cotton wool and bandage it, making a bag to go over all which the goat cannot easily remove. Apply the solution of carbolic acid and glycerine once a day.

There is sometimes a form of foot trouble, known as Fever-in-the-feet, that occurs in cases of mastitis. This is eased by the application of linseed or bran poultices; but it is, of course, the health of the animal that is at fault, and the condition of the feet improves as the disease is mastered and the general health improves.

Indigestion. In goats, this generally appears in the form of flatulence, the body of the animal often being blown up to a tremendous size. An animal in this state is said to be blown, or suffering from blows, pot-belly, the pod, or hoven. There are various local names. It often occurs in the Spring, when animals gorge themselves on lush greenstuff, and it is frequently found in young kids at weaning time. Bottle-fed kids often suffer from this complaint, through taking their milk too fast. The condition must be relieved as quickly as possible ; but, in addition to this, the cause of the trouble should be sought and remedied.

A well-tried remedy for this complaint is four ounces of linseed oil, with a teaspoonful of turpentine shaken up in it. Another reliable remedy is twenty to thirty drops of chlorodyne, in warm water. Many people use a proprietary pain-reliever sold for this complaint in farm stock, and there are various excellent makes on the market. If the dose for goats is not given, treat as for sheep. Some of these mixtures can, also, be given to kids. Chlorodyne should not be given to young kids, but they can have a few drops of essence of ginger or peppermint, in a little warm water.

Personally, I am a great believer in fresh peppermint, and I always advise goat-keepers to grow a patch. A root can be bought from any of the well-known seedsmen, and it grows rampantly in almost any soil, like garden mint. Put it in some odd patch that you do not use, and it will spread happily all over the place and, in a few years, you will have a regular plantation. Like household mint, it roots from underground stems. It sends down roots, also, from the nodes of trailing stems, like strawberries. Goats of all ages will willingly eat it, both in the fresh and dried state, and it is far better to get the patient to take its dose in the form of a pleasant titbit, instead of having to pour things down its throat. Another advantage of having a peppermint plantation is that you do not have to worry about working out doses. Any goat with mild tummy trouble can be given a handful or two of peppermint to eat, and there is no question of causing trouble by giving too much or giving it to too young a goat. They just eat it for pleasure, and it is surprising how quickly a blown goat will go down to her normal size. Cut and dry any peppermint left standing in the autumn ; and, if you have plenty, cut some in the summer, just before it blooms, as it then has the strongest flavour.

Inflammation of the Eyes. This is generally the result of a cold and is a similar condition to that found in puppies, kittens, etc. If it is taken in hand at once, as soon as any sign of eye trouble is noticed, it is quite easily cured and the sufferer is none the worse for it. The eye should be bathed twice a day with warm water and boracic powder, as for humans, and gently dried with a soft cloth, and a small amount of that old stand-by, Golden Eye Ointment, should be applied. Carefully draw out the lower lid a little, squeeze the ointment on to the inner side of the lid, (but not on to the eye-ball), draw the upper lid slightly over the lower for a moment, and then hold the warm hand over the whole eye, while the ointment melts and spreads itself over the eye. Whatever you do, do not neglect this treatment and let the eye become closed up, or the goat may lose its sight. The eye is often closed in the morning, after a night's sleep, but the bathing and the use of the ointment should cause it to open quickly. If the early stages are neglected, a film will form over the eye, and this will eventually come to a head before dispersing. The treatment is the same, but it has to be carried on much longer.

Mastitis. This complaint is known, also, as Mammitis, Garget, and Caked Bag. It is inflammation of the udder and is very contagious, being spread not only by contact between infected and healthy animals but, also, by flies, by the hands of the milker, and by his garments. Mild cases can be cured, if taken in hand promptly, by quite simple measures ; but, once the disease obtains a firm hold, it is a serious matter and must be treated by a vet. If the disease is not brought into your herd from some outside source, it may even develop from chills or injuries to the udder.

The first signs of mastitis are hardness or lumpiness of the udder. The milk is curdly and tastes distinctly salty. The affected animal should immediately be given a dose of a reliable make of udder drench, and the udder should be massaged with an udder salve. These things are sold for cattle, which are very subject to this disease, and they are equally suitable for goats and, needless to say, much better than old-fashioned remedies, such as lard and camphor. If the udder is very hot, it can be bathed with hot water and boracic, but it must be done very gently and the udder thoroughly dried. Milk the sufferer three or four times a day, taking away as much as possible each time; but be sure to milk gently and carefully, so as not to increase the degree of inflammation. The milking should be done into a pail containing disinfectant and the contents should be immediately emptied down a drain. The milker's hands should be washed with disinfectant and special boots and overalls used, when attending to the sufferer. The goat herself should be strictly isolated and kept in a warm, but well-ventilated, building.

Milk Fever. This is another serious disorder, in which the services of a vet. are essential. The goat shows little sign of illness at first and may even be thought to be just lazy, for the trouble comes on suddenly and without any apparent cause. The goat just seems either unaccountably tired or lazy, wants to keep on lying down, and eventually will not even exert herself to eat. The trouble is sometimes due to calcium deficiency, caused through milking a newly-kidded goat dry. No home treatment will suffice, and the vet. must be obtained immediately, before the sufferer becomes unconscious. Treatment is simple and recovery rapid. Unlike cows, goats are liable to suffer from milk fever some months after they have kidded.

Poisoning. As explained in the chapter on feeding, a goat may often eat definitely poisonous plants without coming to any harm. There are many factors to be considered,—the amount eaten, the amount and variety of other items eaten at the same time, and the stage of growth of the plant in question. But it does sometimes happen that a goat, having eaten some poisonous plant, becomes suddenly and dangerously ill; and it is as well to know what to do for the best, while you are waiting for the vet.

Poisons may be roughly divided into two classes,—those that cause violent pain, probably with diarrhœa and vomiting, and the narcotic poisons, which quickly result in unconsciousness. If the goat is in great pain, the first thing to do is to relieve it, and this can be done by giving thirty drops of chlorodyne or thirty grains of aspirin, in either case given in warm water. Let the aspirin tablets dissolve and then give as a liquid dose. Bryony and rhododendron are poisons of this class. Having eased the pain, you can keep up the strength with stimulants, such as black coffee, strong tea, whisky, brandy, or sal volatile, until the vet. arrives. Sal volatile and alcoholic stimulants should be diluted before being given. Not more than a tablespoonful of alcohol should be given at a time.

With narcotic poisons, such as laburnum and yew, strong stimulants must be given, to prevent the animal going into a coma; and the services of a vet. should be obtained at once, as the stomach pump must be used. If there is some delay in getting a vet., give the sufferer a strong dose of linseed oil and castor oil, and use an enema,—the kind used for humans. Use at least a pint of warm, soapy water. The important things are to keep the goat from going into a coma, by the use of stimulants, and to endeavour to clear out the poisonous matter before it does harm. The stomach pump is the best and quickest method; but do not let precious time slip away, while you are waiting, or you are likely to lose your goat.

In cases of deliberate, malicious poisoning, it is often very difficult to know what to do, since the nature of the poison is probably unknown. All one can do is to send an urgent message for the vet. and treat for pain or imminent coma, whichever seems apparent. One can judge this by the symptoms. If you suspect phosphorus poisoning, through the use of Rodine,—and the smell of this, coming up with the breath, is quite unmistakable,—*do not give oil of any sort.* There is nothing for it but the stomach pump, and it must be used very quickly, or the goat will die in great agony.

Scours. See Diarrhœa.

Self-sucking. This is certainly not a disease. It is a vice, and a very troublesome one at that, and treatment should be applied at once. Taken in its early stages, it can generally be cured by smearing the teats with a mixture of lard and asafœtida. This has such a foul taste and smell that few goats will attempt to suck it off, even to get the milk. But the nuisance of such a measure is that the mixture must be washed off before each milking and then re-applied. If the vice has only recently developed, a few days of having the teats treated with this noxious substance may cure the goat; but, if you have a really persistent case, she is likely to wait for the day when you cease to use it and then start her old tricks again. The only thing to do then is to make her wear a special headstall, which has sharp spikes projecting from the band that goes round the muzzle. They do not prevent the goat from grazing, as the band does not fit tightly, and the spikes lift slightly as the wearer puts her muzzle to the ground; but, if she turns round and tries to grip a teat, the sharp points press into her own flesh, and she gives up the idea of her own accord.

Sore Teats. The teats sometimes become sore, chapped, or cracked. They may even develop small ulcers. Warts on the teats are by no means uncommon. The latter are generally thought to be due to a lack of calcium in the system, but the other conditions often arise from careless management. If the teats become wet during milking, they should be carefully dried. To leave them wet not only encourages these sore conditions, which make the goat restive when being milked, but is, also, liable to result in chills and mastitis. Another form of injury to the teats and udder is caused by scratches and tears from brambles, etc., especially when a goat is in full milk and her udder hangs low. In some cases, the teats almost trail on the ground for the first few days after kidding, and the goat herself is liable to tread on them.

Injuries to teats or udder should be treated promptly, however slight they may seem, for flies fasten upon the smallest opening and aggravate the trouble. Carbolised vaseline, or one of the well-known udder salves, is generally all that is needed. See that the udder is thoroughly dry and clean, and apply after milking, daubing on gently in the case of scratches that have bled, and massaging well into the skin in the case of soreness alone. Small ulcers should be painted with iodine. When there are deep scratches or tears, especially when they are aggravated by flies, it is best to wash them with warm water and a mild disinfectant and then paint over with collodion. This makes a film which gives the wound a chance to heal beneath it.

Warts are not, in themselves, harmful, but, if there is one in such a place that it makes the goat restive when being milked, it should be removed. The time for doing this is when the goat is dry. Most warts can be removed by applying a ligature of silk, tied tightly round the neck of the wart, close to the flesh. This cuts off the blood supply, and the wart dies and eventually falls off, of its own accord. When there is no neck, caustic potash must be used, dabbing the wart each day, until it has been burnt away. Be careful only to dab the horny top of the wart and not to burn the tender flesh.

Worms. Goats, in common with other animals, suffer from internal worms and must be treated at times. I do not, however, recommend frequent and regular dosing. Worms become a mania with some people, and worming becomes part of a regular routine. This I consider a mistake. Unless a goat shows definite signs of having worms, there is no sense in dosing her. A vermicide or vermifuge, to be effective, must, of necessity, be pretty drastic ; and what upsets the worms is likely to cause no little inconvenience to the goat. Therefore, do not dose for worms unless you suspect their presence ; and do not, at any time, dose a well goat to keep it well. Leave well alone. Most animals harbour internal parasites without ill effect. It is only when they become numerous that they cause trouble; and this state of affairs is often due, in the first place, to letting the animals get out of condition, so that the worms get the upper hand.

If a goat is suffering from worms to the detriment of her health, it will be seen that she is in poor condition. The coat has a dull, "staring" look, her breath smells unpleasant, and she will have attacks of diarrhœa. Worms will probably be seen among the fæces. There are a number of worm remedies on the market, and, when you find one that satisfies you, it is best to stick to it. Some people swear by one thing and some by another. Personally, I prefer to give goats worm medicine in tablet form. Two tablets are given as a dose, and they can be given in stale bread. Let the goat miss her evening hay and fast through the night. Then give her the tablets, first thing in the morning. Take a few pieces of stale bread to her and give her them as tit-bits. In two of them place the tablets, one in each piece. I have never known a goat refuse them, when given in this way. About three hours later, give the goat a feed of hay and a drink of water. At mid-day, she can go back to her normal diet. The worms should be passed from twelve to thirty-six hours later, and any that are seen should be burnt. Worm medicines should always be obtained fresh for the purpose, and the dosage depends on the weight of the goat. Whether you use a proprietary worm medicine or whether you have it made up for you by the chemist or the vet., remember that it is the weight of the goat in question that must be considered. Never worm an in-kid goat. The milk yield may drop when a goat is wormed, but it soon picks up again.

This covers most of the illnesses found among goats, in this country, and, with good management and reasonably good luck, you should experience few of them in your herd. Tuberculosis, that scourge of dairy cattle, is practically unknown among goats in this country, and the same applies to "Malta Fever". Probably the most frequent cause of losses among goats is the fading out that comes through running them on goat-sick land, of which details are given in Chapter Five, on The Use and Abuse of Pasture; and this is sheer bad management and calls for fresh land, good food, and a tonic.

When you have to drench a goat, get your dose ready, preferably in a small teapot, though a bottle can be used. Back the goat into a corner and either put yourself astride her or lean against her side, holding her firmly against the wall. Put your left arm across her neck, and let your left hand go under her chin and hold up her head, with your left thumb pressing into her cheek behind the back teeth. Then insert the spout of the teapot on the right side, touch the roof of the mouth with it, (to warn the goat that something is coming), and then pour slowly. Keep the head up, but not so high that she has difficulty in swallowing; and let her have a breather now and then, if the dose is large. You do not need to let her go. In fact, it is better not to do so. Hold her firmly, and be businesslike about it. Do not leave her room to shift about and upset herself—and you! But you can give her a short rest from swallowing, while you still keep her in position for the next instalment ; and it is better to do this than to rush her and risk having the medicine go the wrong way.

If you have a number of goats, you might like to invest in what is known as a " Pig Gag ". These are sold for dosing pigs and are very handy for goats. A new one recently brought out by a well-known firm of chemists is particularly useful. There is a wooden piece that goes across inside the mouth and holds it open, and, in the centre of this piece, there is a hole to take a metal measure containing the dose. The measure is marked for doses, like a medicine bottle. I like this gag ; but I do not like the kind that merely consist of a wooden gag with a hole in it, as they leave the attendant the chance of pouring in a large dose too quickly.

There is, also, a very useful pill carrier, made of metal, with a rubber tube to go partly down the throat. The pill is placed in position, the tube put well into the mouth and a spring pressed, after the manner of a syringe. The pill goes well down and cannot be coughed up again.

When sending a goat away by rail, it is always best to rug her, as explained in this chapter on treating sick goats. A folded sack across the back, tied as described, keeps the goat warm, when standing about in draughty places at railway stations, and saves many a chill. Personally, I think the railway people go out of their way to treat animals well, but a certain amount of standing about is inevitable, especially when there are changes ; and a travelling goat is likely to be in a nervous state and so more liable to contract a chill. Remember to put at least two labels on a goat and to put them where she cannot eat them. Many a goat has gone up and down the country, looking for an owner, because she has eaten her labels ! A square of three-ply wood makes an excellent goat label.

CHAPTER FIFTEEN

WHAT DOCTORS SAY ABOUT GOATS' MILK

VERY high claims are made regarding the value of goats' milk, some of them so extravagant that people are often inclined to question their authenticity and ask if goats' milk really has any advantage over cows' milk. It certainly has,—in fact, it has several advantages,—but goat lovers should beware of making claims of which they have not irrefutable evidence. We may well believe that goats' milk saved a life in a certain instance, and our belief may be founded on fact ; but, without something more authentic than our belief behind it, (however well founded), we may find our claims dismissed by many people, on the grounds that they are unreasonable, or even fantastic.

There certainly have been some amazing cases reported of lives being saved and complaints being cured, just by the use of goats' milk. " What is there," people ask, " about goats' milk that can cure, for instance, stomach ulcers and tuberculosis ? " It sounds too good to be true,—too simple, as Naaman evidently thought, when told to bathe in the River Jordan ! Whether or not these, and other complaints, are actually cured by the use of goats' milk it is hardly for anyone outside the medical profession to say ; but I propose to end this book with a chapter about the value and virtues of goats' milk, as they strike the practical man and woman, with extracts from statements that have actually been made by doctors.

The advantages of goats' milk, as opposed to cows' milk, as they strike the practical men and women who have used it, are four-fold. Firstly, it is healthier. We believe this, because tuberculosis among goats is practically unknown in this country, and because the goat is cleaner and more fastidious in her habits.

Secondly, it is, or should be, cleaner, because the fæces of the goat are produced in solid form and soiled udders are rarely seen. Thirdly, the milk is richer, having a higher percentage of butter-fat. Fourthly, we know, from results, that it is more digestible. The cream globules are smaller than is the case with cows' milk, and they are more evenly distributed in the milk and do not rise so readily.

This easier digestibility of goats' milk is very important, for extra richness would be a doubtful benefit if it were hard to digest. But goats' milk can be kept down by many delicate people, both infants and adults, who cannot digest cows' milk; and the author has used goats' milk, for many years, for the rearing of valuable, pedigree dogs and cats. It has been given to the expectant and nursing mothers and used for weaning the youngsters, and there is something about goats' milk that gives results not obtained with cows' milk. The puppies and kittens thrive well and do not have the setback so often seen at weaning time. This I attribute to the easier digestibility of goats' milk, since it is more than can be put down to the greater richness alone.

The following are the analyses of goats' milk and cows' milk, being averages taken from the composition of thousands of samples and given by F. Knowles, F.I.C., Honorary Analyst to The British Goat Society:—

	Goats' Milk.	*Cows' Milk.*
Fat	4·5%	3·67%
Milk Sugar	4.08%	4·78%
Casein	2.47%	2.63%
Albumin and Globulin	·43%	.6%
Non - Protein Nitrogenous Compounds	·44%	·19%
Lime	·19%	·18%
Phosphoric Acid	·27%	·23%
Total Mineral Constituents	·79%	·73%

As will be seen, goats' milk is richer in fat than cows' milk but lower in sugar. The sugar deficiency need not, however, cause any worry, as fats have two-and-a-half times the energy value of carbo-hydrates, as represented by sugar. As regards vitamins, it is a mistake to suppose that these must be deficient because there is no natural colouring matter in goats' milk. There is no reason to suppose that goats' milk is inferior to cows' milk in the matter of vitamins; and, as regards the anti-scorbutic and anti-rachitic vitamins, C and D, definite evidence has been produced that goats' milk is superior to cows' milk.

Discussing the nutritive value of goats' milk, and the fact that it is often retained and digested by infants who cannot keep down cows' milk, Mr. Knowles points out that the curd formed in the stomach is softer and of a more open texture than that formed when cows' milk is taken. The more open texture of the curd, he says, makes it easier for the digestive juices to penetrate it. Also, the fat presents a bigger area for the action of the digestive juices, owing to the smaller size of the fat globules. This makes for easier digestion of both fat and casein. In addition to these things, Mr. Knowles points out the value to digestion of the greater quantities of salts of sodium and potassium, especially the citrates, and draws attention to the common practice of adding sodium citrate when cows' milk is used for feeding infants.

Dr. C. Stanley Steavenson, M.B., who says he would never dare to bring up a child of his own on cows' milk, for fear of tubercle, speaks of very many cases of children who could not digest cows' milk but throve the moment they were put on goats' milk and grew into really healthy children. He, also, gives the case of a consumptive woman who could not take cows' milk, on account of persistent vomiting. As soon as she was put on goats' milk, the sickness ceased and she immediately started to improve.

Melbourne Thomas, M.B., B.S.(London), F.R.C.S.

(Ed.), Llwynypia Hospital, Wales, says that goats' milk may be taken to be tubercle-free, and he sees no reason why it should be pasteurised or boiled, so long as it is collected under hygienic conditions and given reasonably new. Such milk, he says, is given to baby and child patients at the hospital, admitted for such conditions as abdominal tuberculosis, malnutrition, and rickets.

Elizabeth Williams, M.B., Ch.B., D.P.H., (Rochdale Corporation Infant Welfare Centre), states that goats' milk can be more easily modified to the proportions of human milk. The protein, she says, seems more akin to that of human milk. It is more easily digested and assimilated than cows' milk, probably because the rate of growth of a human baby is more like that of a kid than a calf. Seldom, she adds, are undigested curds found in the stools, when goats' milk is given.

The late Dr. B. D. Z. Wright, M.R.C.S., L.R.C.P., late Chairman of Norfolk and Norwich Hospital, and himself a noted goat breeder, gave many instances of cases where patients had benefited from goats' milk. From among them, I quote the following: —

"Baby H., aged 3 weeks. This child had an operation for Pyloric stenosis. Difficulty in retaining food afterwards. Almata, Benger's, cows' milk, (diluted), were tried, but nothing suited. Goats' milk eventually proved its value, being retained and digested. The baby gained steadily ½lb. per week from the commencement of this diet."

"G. A., aged 42. Had a long sickness, (eleven months), from Encephalitis Lethargica, commonly called Sleepy Sickness. Vomiting a marked feature, with much loss of flesh, from 9 st. 6 lbs. to 6 st. 5 lbs. I suggested goats' milk, but the idea that the milk was 'strong' was in the patient's mind. However, she consented to try it and was surprised to find it quite palatable. In a few days, was able to take two pints a day, which were retained. Patient at once began to gain weight and steadily increased. Continued on goats' milk for three months."

"B. T. A., aged 15. Operation for Appendicitis. Tubercular mesenteric glands found. Cows' milk forbidden. Medical attendant writes:—'Benefited greatly from goats' milk, gaining $1\frac{1}{2}$ lbs. a week from the diet'."

"Baby C., aged 5 months. A poor, puny, little thing, much smaller than its nephew, who was a month younger and was breast-fed. Medical attendant advised goats' milk. A goat was purchased and baby fed on the milk. She has done so well on this diet that she has now beaten her nephew in size and weight."

"Baby G., aged 7 months. Food all disagreed with child, till goats' milk was used. Vomiting all ceased, and he has done well on goats' milk."

"W., baby twins. Small and flabby, often sick and not growing. Did well on goats' milk."

"Mrs. H.'s baby. Three weeks old and weakly. Breast-fed. Contracted Broncho-pneumonia and was unable to suck from weakness and was gasping for breath. I was asked to see this child, as death seemed imminent, and the medical attendant was out. Baby had kept nothing down for forty-eight hours, was blue and sweating, pulse too rapid and thin to count, respiration very rapid. I tried goats' milk and brandy. Baby retained this, in gradually increased quantity of milk and diminished quantity of brandy, and recovered, though at death's door for forty-eight hours."

These are some truly remarkable cases, vouched for by medical men, and we can hardly dismiss their testimony as unauthenticated.

Still more recently, we have a report from a Dermatologist, H. Smith Wallace, M.B., who says the medical profession has long recognised the value of goats' milk in certain conditions where people cannot digest cows' milk and proceeds to give his experience of the very satisfactory results from prescribing goats' milk for infants suffering from allergic eczema. This complaint, he says, is due to sensitization of the skin by proteins in the blood-stream and generally comes on when the infant is weaned and put on to cows' milk. In his personal experience, many babies have benefited by being put on to goats' milk and have been completely cleared in a few months, or even a few weeks, by this measure alone. In some cases, the children develop asthma and continue with alternate attacks of asthma and eczema for some years, but they, too, greatly benefit by taking goats' milk. This milk, he says, though chemically much the same as cows' milk, has a softer curd and is more easily digested. Further, it does not seem to contain the same proteins to which the child is allergic, and, consequently, the eczematous reaction in the skin does not occur.

My thanks and acknowledgements are due to the British Goat Society, for permission to reproduce much of the material given in this last chapter, and to the various authorities mentioned by name.

Notes on Illustrations

My thanks and acknowledgements are due to the following breeders and firms who so kindly lent me photographs for reproduction: —

Mrs. Arthur Abbey, Downe Hall, Roydon, Essex.
Messrs. Boots Pure Drug Co., Ltd., Nottingham.
The Dairy Supply Co., Ltd., Cumberland Avenue, Park Royal, London, N.W.10.
Miss J. F. Fawcett, Northmoor Dairies, North Moor, Easingwold, Yorks.
Messrs. Gregory, Ltd., Liverpool.
The Misses M. and V. Window Harrison, Yew Tree Poultry and Goat Farm, North Weald, Essex.
Miss K. Pelly, Theydon Place, Epping.
Ernest S. Smith, Esq., The Rothesmith Herd, Welham Lodge Farm, Moorgate, Rotherham, Sheffield, Yorks.
T. A. Urie, Esq., Broomfield Nursery, Broomfield Drive, Reddish, Stockport.

All the animals shown are worthy specimens of the breeds they represent, and some of them are famous winners, and they show the reader just what a good goat of each breed should be like. When a star or a Q star is shown after a goat's name, it means she has gained a certain award for high production and, in some cases, extra high quality of milk. The same applies when daggers and section marks are shown against the name of a male goat. They are indications of high production on the part of his parents and grandparents.

The Misses Window Harrison and Mr. Urie follow a principle I so often advise and, in fact, carry out myself, that of combining goats with other forms of food production. Other small livestock, gardening, and fruit growing, all work in with each other, and much otherwise inevitable waste is avoided.

A neuter goat, such as Mr. Urie's Dinkie, is particularly useful on such a place as a poultry farm or smallholding, for they are strong and sturdy and yet small enough to take a little trap or cart through a poultry-farm gate. Dinkie was bred and trained by Mr. Urie and was three-and--a-half years old at the time the photograph was taken, and stood approximately three-and-a-half feet high at the shoulder. Such a goat could pull two or three cwts., with a good cart, and could be taken where a larger animal could not possibly go.

INDEX